Urban Geography

Urban geography has reflected the changing nature of geography over the course of the twentieth century and been characterised by a variety of changing perspectives including ecological, new urban economic, feminist and post-modern.

Urban Geography introduces both 'traditional' and contemporary approaches and perspectives in urban geography to explore the activities in global cities, governments and institutions in forming and changing urban landscapes. Placing the European and North American city within a changing set of global economic and urban systems, the book examines the ways in which these systems impact on the lives and the geographies of cities. The rise of the 'post-industrial' entrepreneurial city, resistance of local cultures and political formations, and recent problems that have affected city areas, including homelessness and the edge-city phenomena, are explored.

With the emergence of a new geography of world urbanisation, *Urban Geography* examines the new urban crisis and the challenges for urban governance at both central and local levels.

Tim Hall is a Lecturer in Human Geography at the Cheltenham and Gloucester College of Higher Education.

Routledge Contemporary Human Geography Series

Series Editors:
David Bell and **Stephen Wynn Williams**, Staffordshire University

This new series of 12 texts offers stimulating introductions to the core subdisciplines of human geography. Building between 'traditional' approaches to subdisciplinary studies and contemporary treatments of these same issues, these concise introductions respond particularly to the new demands of modular courses. Uniformly designed, with a focus on student-friendly features, these books will form a coherent series which is up-to-date and reliable.

Forthcoming Titles:

Techniques in Human Geography

Rural Geography

Political Geography

Historical Geography

Cultural Geography

Theory and Philosophy

Development Geography

Tourism Geography

Transport, Communications & Technology Geography

Routledge Contemporary Human Geography

Urban Geography

Tim Hall

London and New York

First published 1998
by Routledge
11 New Fetter Lane, London EC4P 4EE

Simultaneously published in the USA and Canada
by Routledge
29 West 35th Street, New York, NY 10001

Typeset in Times and Franklin Gothic by Keystroke, Jacaranda Lodge,
Wolverhampton
Printed and bound in Great Britain by Biddles Ltd, Guildford and King's Lynn

British Library Cataloguing in Publication Data
A catalogue record for this book is available from the British Library

Library of Congress Cataloging in Publication Data
Hall, Tim
    Urban geography / Tim Hall.
        p. cm. — (Routledge contemporary human geography series)
    Includes bibliographical references and index.
    1. Urban geography. I. Title. II. Series.
    GF125.H35 1998
    910'.9173'2—dc21   97–15274

ISBN 0–415–14084–6
        0–415–14085–4 (pbk)

For Cathy

Contents

Plates

Figures

Tables

Case studies

Acknowledgements

Writing a book, like winning an Oscar, involves many people. I would like to thank all of those who contributed to the production of this text. Colleagues at Cheltenham and Gloucester College and beyond contributed ideas and encouragement; of those I would especially like to thank Heather Barrett with whom I taught an Urban Geography course at Worcester College of Higher Education in 1994–5. Many of the ideas in the book had their origin during that period. I would also like to thank the series editors and those at Routledge who constantly kept me on my toes during the production of the manuscript, especially for their patience when my promised emails, faxes and phone calls constantly failed to materialise. The production of the excellent artwork was by Kathryn Sharp and John Ryan.

Much of this book was written in Cheltenham in 1995–6. I have to thank Cathy for the constant encouragement, endless cups of coffee and her excellent job amending a very rough and ready first draft. Thanks are also due to Goldie, Black Grape and the Chemical Brothers for providing the soundtrack to many a late night at the word processor.

Many thanks to the following for granting permissions for the figures, tables and plates included in this volume: Abbey National Building Society; Addison Wesley Longman; American Academy of Political and Social Science; The Art of Change; Birmingham City Council; The Black Country Living Museum; Blackwell Publishers; Butterworth Heinemann; Carfax Publishers; Paul Chapman Publishing; Department of the Environment, Transport and the Regions; *Employment Gazette*; *Financial Times*; Geographical Association; The *Independent*; NEC Group Limited; Pine Forge Press; Routledge; John Rennie Short for supplying the Syracuse plates; John Wiley and Sons.

1 New cities, new urban geographies

● **New cities and new urban geographies**
● **Different types of city**
● **Geographical characteristics of the industrial city**
● **'Post-industrial' cities – the case of Los Angeles**

New cities, new urban geographies?

The only consistent thing about cities is that they are always changing. Classifying and understanding the processes of urban change present problems for geographers and others studying the city. Cities, since their inception, have always demonstrated gradual, piecemeal change through processes of accretion, addition or demolition. This type of change can be regarded as largely cosmetic and the underlying processes of urbanisation and the overall structure of the city remain largely unaltered. However, at certain periods fundamentally different processes of urbanisation have emerged; the result has been that the rate of urban change has accelerated and new, distinctly different, urban forms have developed. This occurred, for example, with the urbanisation associated with industrialisation in the UK in the nineteenth century.

Geographers have constantly to ask themselves whether the changes they observe are part of the continual process of piecemeal change or whether they are part of more fundamental processes of transformation. Just such a debate has been occupying geographers, sociologists and other social scientists in the latter part of the 1980s and the early 1990s. The issue of whether we are witnessing the emergence of new types of cities has also raised questions about the adequacy and relevance of the geographical models and theories developed in the past to understand cities.

The earlier mention of the industrial revolution raises issues of investigation that shape the themes of this book. Do we need to look at the changes in, not only national, but also the international economy, since the 1970s and ask ourselves whether or not they are as epochal in their extent and significance as those changes now labelled the 'industrial revolution'? The answer to this question is unequivocally yes. There is little doubt that since the early 1970s the world economy has been affected by a number of fundamental changes. The ramifications of these changes have been enormous and have affected, not only the economic life, but also the social, cultural and political lives of nations, regions, communities and individuals. Tracing the links between the changes in the world economy and those in the landscapes, societies, economies, cultures and politics of cities is the main aim of this book.

Visually, the evidence of a fundamental transformation of the processes of urbanisation appears compelling. The signs of significant change are apparent in many urban landscapes of North America, the UK, mainland Europe and many parts of the developing world. Some of the most widely debated of these signs of change have been the enhancement of city centres by extensive redevelopment, the redevelopment of derelict, formerly industrial areas such as factories and docks (see Plate 1.1), the use of industrial and architectural heritage in new commercial and residential developments, the social, economic and environmental upgrading of inner-city neighbourhoods by young, middle-class professionals, the appearance of brand-new 'city-like' settlements on the edges of existing urban areas and the emergence of large areas of poverty and degradation, for example in old inner-city areas and on council housing estates on the edges of numerous towns and cities.

The ways in which recent changes in the urban landscape have been reported in both the popular press and by geographers have tended to support the notion that they represent some form of transformation. Enthusiastic local newspapers frequently present major projects of urban regeneration as signifiers of the transformation of a city's fortunes, ushering in a new era of prosperity in the service, rather than the industrial economy. In a similar manner, the language that academics have used to debate and describe contemporary urban change would suggest that some profound differences in the urbanisation process have emerged. The language that academics have used to describe these changes has included: from industrial to post-industrial, from modern to post-modern, from Fordist to post-Fordist. However, despite apparently compelling visual evidence and the language used to describe change, it

Plate 1.1 *The new face of British urbanisation? London's Docklands*
Source: Ambrose (1994: 63)

is important to try and remain objective and to assess the degree to which these could truly be called a transformation of the urbanisation process and an emergence of new forms of urban settlement.

Some of the questions that emerge from this debate include:

- How significantly has urban form being altered?
- How have these changes varied geographically between different cities?
- How differently does urban life feel now? And for whom?

This book explores these, and other, questions. It will consider the structural, economic, political, social and cultural changes that have affected urbanisation, primarily in the UK, but with reference to mainland Europe and North America.

Different types of cities

Many urban geographers and historians have argued that the cities that we recognise in the 1990s are the product of a long evolutionary process, during which the settlements of 15000 BC gradually evolved into the complex cities of the late twentieth century. This view may seem very appealing. However, it ignores some very important dimensions of contemporary urbanisation. No two cities are identical. They may be broadly similar, but cities have very different landscapes, economies, cultures and societies. This is a reflection of the fact that cities are shaped by a diverse set of processes. The particular set of processes that affect city development depends on a number of factors that are unique to individual cities, such as city size and the nature of its economy, and/or related to wider factors, such as the relationships between networks of cities, the nature of the nation within which they are located and their position within the world economy. The diversity of city types and processes of urbanisation cannot be reduced to a simple, linear evolutionary process. It is preferable to adopt a perspective that recognises this diversity and to think of cities having different roles and positions in the world economy. The trajectory of urban development is bound up with the workings of the world economy and the relationships of individual cities to this (Savage and Warde 1993: 38). The following classification of different types of cities recognises this:

- Third World cities
- Cities in socialist countries
- Global (world) cities
- Older (former) industrial cities
- New industrial districts

(Savage and Warde 1993: 39–40)

This classification is not totally comprehensive, nor should it be applied too rigidly. For example, many cities fall into more than one of the categories listed (Savage and Warde 1993: 40). London is a global city; however, it contains a considerable decaying industrial economy and yet it is surrounded by many new industrial districts. It is often far from easy to determine which category describes a city best. Further, there is a great deal of diversity within each category, especially between Third World cities but even between older industrial cities which might have been based upon different industries. Despite these limitations this classification recognises that urbanisation is different in various parts of the world and for different types of city (Savage and Warde 1993: 38–40). This book focuses primarily on the urban geography of older industrial cities, global cities and new industrial spaces.

Evolution of the industrial city

Cities which were formed, or in large part influenced, by the processes of urbanisation linked to the 'industrial revolution' of the nineteenth century are of relevance to a discussion of contemporary urban geography for a number of reasons. First, such 'industrial' cities constitute a large part of the urban systems of the UK, the USA and Europe. Industrialisation influenced the internal geographies of many cities in these regions as well as the economic, political and physical links between them. These legacies have formed important dimensions of subsequent urbanisation. These cities are variously referred to in debate as 'modern' or 'industrial'.

As well as the obvious importance of these cities to the urban geography of older industrial nations, the industrial city has also had a disproportionate influence on modern urban theory. The distinctive built form of the industrial city became well known in the latter half of the nineteenth century. The dreadful conditions that prevailed in its inner residential areas made it ripe for study by journalists, satirists, social reporters and novelists. Novels which explored the dark sides of these cities include *Sybil: The Two Nations* by Disraeli (1845), *North and South* by Elizabeth Gaskell (1848) and *Hard Times* by Charles Dickens (1854); social reports included Frederick Engels' exploration of Manchester published in *The Condition of the Working Class in England* (1844) and Charles Booth's social survey of London published as *Life and Labour of the People of London* (1889). The horrific images of these texts did much to form the initial middle-class reaction to the industrial city and the lasting perceptions still present in the 1990s.

In North America the link between urban change and the formation of urban theory was even more direct. For much of the early twentieth century the most influential sociology department in North America was at the newly formed University of Chicago. The specialism of this department was urban sociology and it was in Chicago that much of their research was conducted (Ley 1983: 22). Some of the most famous publications of this school included Robert Park, Ernest Burgess and R.D. McKenzie's *The City* (1925), Ernest Burgess's *The Urban Community* (1926), H. Zorbaugh's *The Gold Coast and the Slum* (1929) and H. Hoyt's *One Hundred Years of Land Values in Chicago* (1933). At the time Chicago was a new city; it had grown rapidly and owed much of this growth to industrialisation. Models of urban structure, most famously Burgess's concentric zone model (Figure 1.1) and Hoyt's sector model (Figure 1.2), were based on this research and inevitably reflected the structure of the city and the forces that created it. The influence of these models on Mann's (1965) model of the British city is obvious (Figure 1.3).

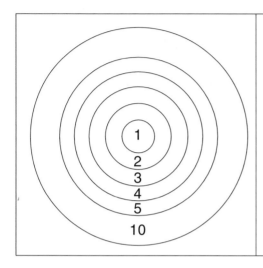

District

1 Central business district
2 Wholesale light manufacturing
3 Low-class residential
4 Medium-class residential
5 High-class residential
6 Heavy manufacturing
7 Outlying business district
8 Residential suburb
9 Industrial suburb
10 Commuters' zone

Figure 1.1 *Burgess's concentric zone model*
Source: Ley (1983: 73) by permission of American Academy of Political and Social Science

District

1 Central business district
2 Wholesale light manufacturing
3 Low-class residential
4 Medium-class residential
5 High-class residential
6 Heavy manufacturing
7 Outlying business district
8 Residential suburb
9 Industrial suburb
10 Commuters' zone

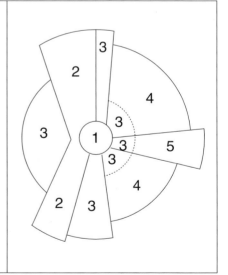

Figure 1.2 *Hoyt's sector model*
Source: Ley (1983: 73) by permission of American Academy of Political and Social Science

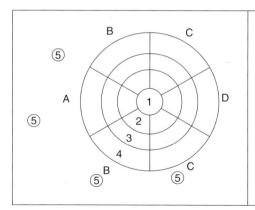

A Middle-Class Sector
B Lower Middle-Class Sector
C Working-Class Sector (and Main Sector of Council Estates)
D Industry and Lowest Working-Class Sector

1 CBD
2 Transitional Zone
3 Zone of small terraced houses in Sectors C and D; larger bye-law housing in Sector B; large old houses in Sector A
4 Post-1918 residential areas, with post-1945 housing on the periphery
5 Commuting-distance 'dormitory' towns

Figure 1.3 *Mann's model of the British city*
Source: Knowles and Waring (1976: 245)

Reactions to the industrial city also shaped a number of nineteenth- and twentieth-century traditions in the fields of political ideology, town planning and literature (Short 1984: 14–16). The physical and intellectual legacy of the industrial city, therefore, is hard to ignore.

The most immediately striking aspects of the industrial city were the extent to which they revolutionised the urban systems of the UK and the USA and the speed at which this took place. In Britain, for example, in 1800 only one city, London, exceeded 100 000 people; however, by 1891 the total was 24 (Ley 1983: 20). Growth figures for individual cities reveal the rapidity of this urbanisation (Table 1.1).

In the USA this process occurred some fifty to sixty years later. Its effects were none the less fundamental. In both the UK and USA the focus of their urban systems shifted away from previous political, religious or mercantile centres towards manufacturing centres. Sources of cheap energy, first water and later coal, acted as strong magnets for this growth, coupled with the accessibility offered by rivers and other waterways such as lakes.

Not only were the size and rapidity of growth of these cities new but so also were the forces that shaped this growth. The growth of the industrial city was tied to the development of

Table 1.1 *Population growth in selected UK cities 1800–1901*

City	Population		
	1800	1851	1901
Birmingham	71 000	265 000	765 000
Manchester	75 000	336 000	645 000
Leeds	53 000	172 000	429 000

Source: Ashworth (1968) quoted in Greed (1993: 67)

the factory system within the emergent form of industrial capitalism. The ability of industry to outbid other uses for land near the centres of cities was fundamental to the generalised form of the industrial city portrayed in contemporary accounts and urban models. The cores of industrial cities remained largely commercial, a land-use that was able to outbid all others for this expensive accessible land. However, surrounding this core was typically a ring of industry, which required a large labour force housed nearby, leading to the development of a ring of working-class housing surrounding this. The regulation of this residential development was non-existent until the late nineteenth century in many industrial cities. Systems of urban government were archaic, based on the parish system, and more suited to the administration of rural areas or small urban areas. Consequently, they were totally overwhelmed by the scale of urbanisation during the nineteenth century. The quality of the housing in this zone was extremely poor and the provision of services and utilities such as running water, lighting and sanitation was frequently nil. These zones became notorious for outbreaks of lawlessness and disease initiating a series of 'moral panics' among the wealthier middle classes. These reactions to the zones of new working-class housing were fundamental to the desire of the middle classes to move away from the centre of cities into the expanding suburbs, a desire facilitated by progressive transport innovation.

Despite the theoretical importance afforded to both Manchester and Chicago they were not typical of the industrial city. Rather, these two cities were the 'shock' cities of industrialisation. They represented extremes where an industrial urban form became most developed and complete. In other cities industrialisation was mediated by local conditions, the legacy, be it physical or otherwise, of earlier rounds of urbanisation. In many cases the transformation of urban form by industrialisation was partial rather than total, mixed in with earlier urban forms, or the form of industrialisation differed from the norm. In Birmingham, for example, initial mass industrialisation was concentrated in small workshops rather than factories. It was not until the early twentieth century that the factory system came to dominate. Further, Birmingham's urban form was disrupted by a complex patchwork of land-ownership patterns which determined land-uses as much as the forces of industrialisation did.

Los Angeles and models of the post-industrial city

> Los Angeles has been a particularly vivid context from which to explore postmodern urbanization in virtually all its dimensions. I have called it the quintessentially postmodern metropolis not because I see it as a 'model' for all other cities to follow or as a doomsday scenario to warn the rest of the world. If there is anything which places Los Angeles in a special position with respect to an understanding of postmodern urbanization processes, it is that comprehensive vividness I referred to, the particular clarity these restructuring processes have taken in this region of Southern California. In part, this is due to the relative absence of residual landscapes derived from preindustrial, mercantile, and nineteenth-century industrial urbanizations. Although founded in 1781, Los Angeles is pre-eminently a twentieth century metropolis. Since 1900, the population of the Los Angeles region has grown by more than 14 million, more than almost any other city in the world.
>
> (Soja 1995: 128)

Los Angeles in the late twentieth century has assumed a position with regard to urban theory comparable to that of Chicago in the early twentieth century. A number of academic and popular accounts have constructed it as an archetype of contemporary and future urbanisation. The shift in the location of the most influential North American school of urban theory symbolises the eclipse of the USA's 'rust-belt' of manufacturing cities by emerging cities of the west coast oriented more to the developing world than to Europe (Savage and Warde 1993: 58). Just as Chicago was home to an influential school of urban sociology, Los Angeles has been home to an influential school of urban studies, the Graduate School of Architecture and Urban Planning at University College Los Angeles, since the early 1970s. This school has included a number of prominent urban theorists who have used Los Angeles as the primary laboratory in which they have researched the emergent processes of urbanisation of the late twentieth century. Important works from this 'California School' have included Allen Scott's *Metropolis: From the Division of Labour to Urban Form* (1988), Ed Soja's *Postmodern Geographies: The Reassertion of Space in Critical Social Theory* (1989) and *Thirdspace: Journeys to Los Angeles and Other Real and Imagined Places* (1996), Mike Davis's *City of Quartz: Excavating the Future in Los Angeles* (1990) and Frederick Jameson's *Postmodernism, or the Cultural Logic of Late Capitalism* (1992). More popular accounts that have explored notions of new processes of urbanisation in cities like Los Angeles, San Francisco, Tokyo and, to a lesser extent, London have included Deyan Sudjic's *The 100 Mile City* (1993). To these can be added a host of articles in academic and professional journals, magazines and newspapers by the authors listed above and many more.

A major theme of many of these works has been the idea of the fragmentation of urban form and its associated economic and social geographies. Namely, that the city is ceasing to exist as a recognisable single, coherent entity; rather it is physically fragmenting as independent cities emerge on the edge of existing metropolises and economically, socially and culturally fragmenting as divisions between different social groups widen to the extent of their becoming broken. The city fragments, according to this logic, into a series of independent settlements, economies, societies and cultures. This is expressed in the idea of the galactic metropolis proposed by Peirce Lewis in 1983 which describes urban form as resembling a series of stars floating in space, rather than a unitary, coherent entity with a definable centre (Knox 1993). This idea of fragmentation was present in two models proposed of the post-industrial city based on research in Los Angeles and Atlanta. These models are the urban realms model (Hartshorn and Muller 1989) which describes a series of separate cities existing within a larger metropolitan area, and Soja's (1989) model of the post-industrial 'global' metropolis (Figure 1.4).

Much of the work of the California School is based on what they argue is a link between changes in the organisation of capitalism (regime of accumulation), and the new industrial sectors and spaces that they throw up, and urban form. These commentators have painted a picture of Los Angeles as a city whose economic and social geographies are based upon new economic growth sectors such as animation (Christopherson and Storper 1986), the motion picture industry (Scott 1988) and hi-tech defence related industry as well as informal, quasi-legal or illegal economic activities of various kinds (Soja 1995).

However, such models can be subject to two sets of criticisms. The first concerns their attempts to link changes in the regime of accumulation with the restructuring of urban space and neighbourhood formation, through the medium of industrial restructuring. To use this as an explanatory basis for neighbourhood formation is very tenuous and narrow and ignores a number of very important issues in neighbourhood formation and reproduction. This reduction to economic causality

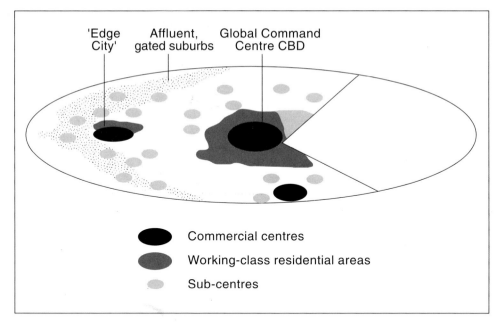

Figure 1.4 *The post-industrial 'global' metropolis*
Source: Graham and Marvin (1996: 334)

reflects a number of other weaknesses in this approach. It pays very
little attention to the micro-social struggles and conflicts that take place
within and between neighbourhoods. This reflects a general neglect of
human action and a focus primarily on the abstract and the structural.
Even within the economic sphere their focus is very narrow as it fails to
pay much attention to the service sector, an important component of any
advanced urban economy (Savage and Warde 1993: 59–61).

. The second set of criticisms are similar to those levelled at models
derived from research in Chicago in the early twentieth century. Put
simply, how typical is Los Angeles of cities in the late twentieth
century? Earlier in this chapter it was pointed out that cities have a
diversity of contrasting histories. Immediately it should be apparent that
to suggest that Los Angeles can provide anything like a general model
of urbanisation is flawed thinking. It would be naive to suggest that
California School commentators do not recognise Los Angeles'
exceptional history. However, as much as a result of the prominence
attached to their work on Los Angeles, the city has assumed a
theoretical primacy within contemporary debates on urban form.
Much of what follows in the book examines the question, implicitly
at least, of the extent to which 'industrial' urban forms have been

superseded by 'post-industrial' forms. The question of whether Los Angeles represents a common urban future is one for the reader to actively address.

The most profitable application of the work of the California School of urban geography lies not in the issue of modelling overall urban forms, but rather, the recognition they afford to emergent trends in processes shaping urban landscapes, economies and cultures, which they discerned first, and with most clarity, in Los Angeles. These trends include the rise of new flexible forms of economic organisation and production; the increasing influence of an interconnected global economy and the crucial role played by cities within this; the appearance of 'edge-cities'; the heightening of various forms of economic, social and cultural inequality within cities which are expressed in new patterns of spatial segregation; the rise of 'paranoid' or carceral architecture based on protection, surveillance and exclusion and finally the increasing presence of simulation within urban landscapes, imaginations of alternative cities to the 'dreadful reality' of actual cities, for example theme parks, themed, policed shopping malls, and more subtle forms of simulation that invade everyday life. Ed Soja summarised these trends as 'six geographies of restructuring' (Soja 1995: 129–37). While these should not be uncritically imported as a blue-print for post-industrial urban change, they signal processes discernible to varying extents in many urban localities.

New cities?

Clearly cities, and the processes that produce them, have changed and will continue to change. However, it is important to try and understand the significance of these changes. Are we witnessing the emergence of new types of cities produced by new forms of urbanisation or are cities remaining fundamentally unaltered except for some cosmetic changes? Or is the situation somewhere between the two?

On the transformation side it is clear that the changing nature of international capitalism (which is explored in more detail later) has had a major impact on the relations between cities and the internal geography of the city. However, while an undoubted resorting has occurred the overall structure of the city is still recognisably modern, rather than post-modern. The structure of the city that has evolved over the course of the twentieth century and been structured by industrial capitalism and national systems of town planning has far from been completely overwritten since the early 1970s.

Rather than a new city emerging it would be truer to say that 'newer' formations have begun to appear within a more traditional structure. The extent to which the newer elements will come to dominate is difficult to determine. The city is currently experiencing more than just piecemeal change; however, it is some way short of being completely transformed (Cooke 1990: 341; Soja 1995: 126).

A framework for examining urban change

The urban landscape does not simply appear overnight, rather like a movie-lot springing up on vacant land, but has to be *produced*. A large number of actors are involved in this production: architects, designers, builders, property developers and construction workers among others. As well as those directly engaged in the physical production of the urban landscape there are a similarly large number less directly engaged but none the less as vital. These include investors in the built environment. The composition and organisation of these groups has changed significantly since the mid-1980s. It has also come to include new and increasingly influential players such as pension and finance companies (Knox 1991; 1993).

The technology employed in the production of the built environment has always profoundly affected its form. The development of innovative architectural forms, such as the skyscraper in the early twentieth century, was the product of the coalition of a number of important factors, not least the development of both building and communications technology. The transformation of the technologies of building production, for example with the employment of computer technology, has profoundly affected the production of the built environment and its form. Each of these components of the production of the built environment is examined throughout the book (Knox 1993).

The urban landscape is not produced without constraint. It cannot simply be thrown up at the whim of a developer or architect. Rather, the built environment is subject to strict control by the state through the planning system. Not only is the urban landscape produced but also it is *regulated*. The regulatory action of the state is an important influence on the development and the direction of physical change in the urban landscape

(Greed 1993). Chapter 5 considers significant recent changes in the regulatory actions of the local and national state in urban development and planning.

The urban landscape is not produced only to be looked at, although its appearance is a vital component; it is primarily built to be used. The urban landscape serves a vast variety of users, be they residential, commercial, industrial, retail or leisure users. Not only is the city produced and regulated but also it is *consumed*. The composition of these groups of consumers and their needs, wants, tastes and ability to consume will affect profoundly what is built for them. Understanding the relationships that surround the changing urban landscape requires that this dimension is also taken into account (Harvey 1989a: 77). Again the consumption dynamics surrounding the urban environment have displayed important changes since the mid-1970s.

The framework outlined above has so far concentrated on the physical components of the city, the production, regulation and consumption of the built environment. However, this is not the sole concern of this book; rather it provides both a fulcrum and a point of departure from which to examine many more aspects of urban life. The early part of the book focuses on factors affecting the production and the regulation of urban change, landscape, society and economy. These are brought together in Chapters 6 and 7. While the focus in these chapters is on changing urban landscapes and images of cities they highlight the ways in which these are structured within certain germane economic, political, social and cultural contexts. Chapters 8 and 9 consider how these changes, in turn, structure new urban social and cultural geographies. In doing so it considers the economic, political, social and cultural impacts of recent urban change. The book concludes with a chapter that speculates how these changes might affect urban development into the twenty-first century.

New urban geographies?

Geography and related disciplines have a long history of models and theories formulated to understand the city and the processes of urbanisation. These have stemmed from several, often contradictory perspectives. Time has witnessed the continual replacement of successive schools of thought with alternatives.

The inadequacy of individual urban models and theories has been continually exposed. Either they have been shown to have been theoretically weak, partial or that cities have changed to such a degree that the models have rapidly become redundant. The question of the adequacy of previous models of the city is again relevant, given the transitory nature of contemporary urbanisation. Chapter 2 briefly reviews the history of urban theory in the twentieth century and considers the question: are existing ways of understanding the city becoming redundant?

Further reading

Good descriptions of the evolution of the industrial city can be found in:

Carter, H. (1983) *An Introduction to Urban Historical Geography* London: Edward Arnold.

Dennis, R. (1984) *English Industrial Cities of the Nineteenth Century* Cambridge: Cambridge University Press.

Mumford, L. (1991) *The City in History* Harmondsworth: Penguin.

A useful introduction to many aspects of the 'newer' cities mentioned above can be found in the following book. Think about how the observations of the authors in these books apply to UK cities.

Knox, P.L. (ed.) (1993) *The Restless Urban Landscape* Englewood Cliffs, NJ: Prentice-Hall.

The best way to get a flavour of the urban theory emanating from the California School is to sample their writings about Los Angeles. Try Mike Davis's very readable account and the two chapters in Ed Soja's book which specifically analyse Los Angeles' urbanisation.

Davis, M. (1990) *City of Quartz: Excavating the Future in Los Angeles* London: Verso.

Soja, E. (1989) *Postmodern Geographies: The Reassertion of Space in Critical Social Theory* London: Verso (Chapters 8 and 9).

A broad discussion of that process of cultural change known as post-modernism with some reference to the city can be found in:

Harvey, D. (1989) *The Condition of Postmodernity* Oxford: Blackwell.

A good guide to the rich and varied history of urban thought can be found in:

LeGates, R.T. and Stout, F. (eds) (1996) *The City Reader* London: Routledge.

2 Changing approaches in urban geography

- Definitions of urban settlements
- The evolutions of urban geography
- Contemporary urban theory

What is urban?

What is urban? While this may seem a good place to start it may also appear a stupid question. Most people know an urban area when they see one. However, the point of the question was not recognition but definition. How are we able to define 'urban'? What, if anything, is uniquely urban? What makes urban areas qualitatively different from other types of areas?

It is relatively easy to classify which areas are urban and which areas are not urban. There are a host of indicators available. These include population size, population density, number and range of services or employment profiles. Some sociologists have even claimed to have recognised distinctly urban and non-urban lifestyles, yet these have been severely criticised (Glass 1989; Savage and Warde 1993: 2). However, classifications such as these are descriptive classifications. While they are able to describe what characteristics are present in urban areas they are unable to isolate and identify anything that is unique to urban areas and which can therefore be used as a basis of a definition of 'what is urban'. These indicators identify differences of degree not type.

For example, shops are found in both urban and non-urban areas. While the shops found in urban areas might be larger, more specialised, of a greater number and of a greater range than those found in non-urban areas, there is nothing exclusively urban about shops. The observable differences are differences of degree not fundamental differences of type. If a classification of urban and non-urban areas was to be devised on the basis of the shopping facilities found in each area, as indeed it has been, the decision of where to draw the boundary between urban and non-urban is not natural, or necessarily even obvious, it is a human decision. A classification such as this reflects nothing more than human opinion, not natural law. Indeed it is the failing of many a classification that the two are often confused.

This fundamental uncertainty in classifications is reflected in the fact that classifications devised in different countries differ so much. In Scandinavia, for example, settlements as small as 300 people might be classified as urban, while in Japan a settlement must have 30 000 people to be classed as urban.

The question 'what is urban?', therefore, is resolved not by asking how many people, shops, offices, etc. make a city, but if the processes which form, transform, operate upon and operate within cities are unique to cities, or whether they are common to both urban and non-urban areas. The answer to this question is no. These processes are not unique to urban areas. The same processes form and transform non-urban areas as form and transform urban areas. While we may find a greater number or range of certain things in urban areas they are not fundamentally different to non-urban areas in that they are the product of common social processes (Glass 1989: 56; Savage and Warde 1993: 2). Poverty, for example, considered a major characteristic of urban, particularly inner-urban, areas is found to an equal, in many cases greater, degree in rural areas. Despite its presence in urban areas, poverty (like so many other things) is a social rather than an urban problem.

The apparently distinctive characteristics of urban areas are not, in fact, distinctive, only different. These differences merely reflect that social processes produce different outcomes in different areas under different local conditions. Given this, 'urban' should not be considered a natural, distinctive type of place, merely a convenient label used by geographers and other social scientists to provide a spatial focus for their work.

Having said this, 'urban' is a meaningful label, for a number of reasons. For example, geographers continue to organise research and publications around themes like urban and rural. Therefore, libraries use 'urban' as a classification under which they file books and journals. Also categories such as urban and rural are reflected in the ways societies, organisations and governments organise and administer themselves. In the UK, examples include the Council for the Protection of *Rural* England, *Urban* Development Corporations, The *Countryside* Commission, The *City* Challenge Programme, and so on. While 'urban' is not a natural, distinctive type of place, merely a convenient label, it is one that reflects social and academic perception and therefore forms a meaningful theme of inquiry.

What is urban geography?

Urban geography is, largely, what urban geographers do. While this might not be a very helpful point it does reflect the subject's lack of precise definition. However, it is possible to recognise a number of concerns common to many urban geographers. These concerns can be summarised as being of three types. *Descriptive concerns* involve the recognition and description of the internal structure of urban areas and the processes operating within them or the relations between urban areas. *Interpretative concerns* involve the examination of the different ways in which people understand and react to these patterns and processes and the bases that these interpretations provide for human action. *Explanatory concerns* seek to elucidate the origins of these patterns and processes. This involves an examination of both general social processes and their different manifestations under particular local circumstances (Short 1984).

However, urban geography is nothing if not dynamic. The emphasis placed upon these concerns has shifted significantly over the course of the twentieth century as urban geography has developed as an academic discipline.

Urban geography has been characterised, over the course of its history, by a series of radical shifts in the way geographers have gone about investigating the urban. This is a reflection of radical shifts in the philosophies that underpinned these approaches. Each of these approaches has been characterised by a very different emphasis on the approaches outlined above. The following section briefly reviews the main approaches in urban geography in the twentieth century.

Changing approaches in urban geography

Early approaches

Site and situation

Studies from the early twentieth century were concerned primarily with the physical characteristics as the determining factor in the location and development of settlements. This concern has been long superseded in all but historical and some rural studies as cities have grown in both size and complexity. Original location factors have tended to be overridden by the scale of subsequent urbanisation or have greatly declined in importance as the form and function of urban areas have changed.

Urban morphology

This was an important root of urban geography. It developed particularly strongly in German universities in the early twentieth century. It was primarily a descriptive approach that sought to understand urban development through examination of the phases of growth of urban areas. Using evidence from buildings and the size of building plots, it aimed to classify urban areas according to their phases of growth. While this approach came in for some heavy criticism in the 1950s and 1960s as more scientific approaches came to dominate the subject and the social sciences generally, it made something of a limited comeback in the 1980s. Recent work has concentrated on the roles of architects, planners and other urban managers in the production of the form and design of urban areas (see for example Whitehand and Larkham 1992).

Modern approaches

The two approaches outlined above were associated primarily with the infancy of urban geography. A greater diversity and maturity was evident in the approaches that came to dominate in the post-1950 period. The most significant of these and their main critiques are outlined below.

Despite their obvious differences, the models outlined demonstrate some similarities. Basically they have all sought to examine the ways in which urban patterns and processes are the outcome of the combination of human choice and action and wider social processes which place constraints upon this human action. Consequently the approaches that follow have all aimed to explore three things. First, they have all considered the ways in which humans make choices about a variety of things (where to shop, where to live, where to build, how to build, etc.) and the ways in which these decisions might influence urban patterns and processes. Second, they have all explored the constraints that might impinge upon this human choice and the ways that this constraint might influence urbanisation. Finally, they have considered the outcomes of the relationship between choice and constraint. They have considered which is the dominant side of the relationship and the ways in which urban development is the outcome of the combination of choice and constraint. What distinguishes each of the following approaches is the relative importance they place on choice and constraint and the ways in which they believe each to operate. Choice and constraint is a dominant theme of urban geography in the post-1950 period.

Positivist approaches

Although the positive philosophy dates back to the 1820s it significantly influenced urban geography only from the 1950s. This reflected the impact of scientific approaches upon the social sciences generally and the increasing capacities of computers allowing the manipulation of ever more complex statistical data-sets.

The positive philosophy is based upon the belief that human behaviour is determined by universal laws and displays fundamental regularities. The aim of positive approaches was to uncover these universal laws and the ways in which they produce observable geographical patterns. Positive approaches can be subdivided into two types – ecological approaches and neo-classical approaches.

Ecological approaches were based upon the belief that human behaviour is determined by ecological principles, namely that the most powerful groups, however this was defined (usually in terms of their income), would obtain the most advantageous position in a given space, the best residential location for example. This school of urban geography dates back to the Chicago School of sociology from the 1920s, and their contributions include Burgess's concentric zone model and Hoyt's sector model of land use (see Figures 1.1, 1.2 and 1.3).

The ecological approach developed during the 1960s in that the models were refined with the increasing sophistication of computers. Despite this, they were able to offer little more than descriptive insights and by the 1970s were being criticised for their failure to say anything about the growing problems seen in cities; they started to become superseded by other approaches (Ley 1983).

An interesting application of these models was that developed by Mann in 1965 of the British city. The principal innovations of Mann's use of Burgess's and Hoyt's models was its combination of concentric and sectoral residential class patterns. This was based on the recognition that while cities might develop outwards in patterns approximating to concentric zones, frequently income differences are sectoral in pattern. Mann also incorporated the actions of the local authority in the provision of housing, an important aspect of the urban geography of post-war British cities, and the effects of negative industrial externalities on residential location. Mann argued that pollution from industry in the core was dispersed to the east of British cities by prevailing winds, thus degrading this area as a residential location. This created an east–west split in residential class. This was sustained by observation in a number of British cities. Despite these innovations, however, this model has dated poorly and is subject to all of the limitations that characterised this approach.

Neo-classical approaches were also based on the belief that human behaviour was motivated primarily by one thing and was, therefore, predictable. However, they believed that this driving force was rationality. By rationality they argued that each decision was taken with the aim of minimising the costs involved (usually in terms of time and money) and maximising the benefits (again time and money). This type of behaviour was referred to as utility maximisation.

The cities produced by positive models, of both types, were of neat, regular, homogeneous zones. Their very poor approximation to reality was the source of much of the criticism directed at these models and reflected the overly simplistic assumptions upon which they were based and the important factors and motivations they ignored. Their failure to recognise and account for the idiosyncratic and subjective values that motivated much human behaviour was critiqued by behavioural and humanistic approaches that emerged in the 1970s and 1980s. These approaches placed the question of the complexity of human motivation at the centre of their inquiry. Positivist theories were also criticised for their failure to consider adequately the constraints within which human decision-making took place. A body of theories and approaches (given the umbrella title structuralist) which again emerged in the early 1970s sought to redress this imbalance.

Behavioural and humanistic approaches

Both of these approaches developed as criticisms of the failings of the positivist approaches. They were united in their belief that people, and the ways in which they made sense of their environment, should be central to their approach. However, they differed considerably in the ways in which they went about this. Behaviouralist approaches can be regarded as an extension of positivist approaches. They sought to expand positivism's narrow conception of human behaviour and to articulate more richly the values, goals and motivations under-pinning human behaviour. However, despite this they were still concerned with uncovering law-like generalisations in human behaviour. Behavioural approaches sought to examine the ways in which behaviour was influenced by subjective knowledge of the environment.

Humanistic approaches stemmed from a very different philosophical background. They sought to understand the deep, subjective and very complex relationships between individuals, groups, places and landscapes. In a radical departure from the scientific approaches of the 1950s and 1960s, the humanistic approaches brought techniques more associated with the humanities to understand people–environment relationships. This was reflected in the sources they utilised. These included paintings, photographs, films, poems, novels, diaries and biographies. The influence of humanism in urban geography was limited. Most humanistic work was conducted on rural or pre-industrial societies. Humanism developed in urban geography mainly as a critique of the monotonous, soulless landscapes of modern cities. The best developed application of the humanistic perspective in this regard is Edward Relph's *Place and Placelessness* (1976).

Both of these approaches, despite their differences, were less concerned with the production of descriptive models of urban form, and more with the production of interpretative insights into the relationship between people and their environments. However, limitations within the approaches themselves, and the criticisms from structuralists, again to do with their failure to consider the constraints acting upon human decision-making and behaviour, limited their long-term impact on the subject.

Structuralist approaches

Structuralist approaches in the social sciences generally, and in urban geography specifically, can be recognised through their conviction that social relations and spatial relations are either determined, or are in some way influenced, by the imperatives of capitalism as the dominant mode of production. This has led to criticisms that such analysis has failed to adequately account for the role of human action within these relations. Structuralist approaches have been accused of treating humans as mere passive dupes of economic structures. Much development in structuralist urban geography has involved the attempt to try and incorporate the 'structural' and the 'human' dimensions and thus overcome criticisms of 'reductionism'. Structural analysis in urban geography has largely derived from interpretations of the work of Karl Marx. Marx's view of history was of a series of 'modes of production' each of which was characterised by a particular structural relationship between the economic base and the social superstructure. Cruder versions of Marxism present the change in the economic base as controlling and determining change in the social superstructure. While this is a considerable over-simplification, the ways in which this relationship has been interpreted have varied a great deal over time and between different schools of Marxist scholarship.

The neo-Marxist influence on the social sciences in general dates back to the late 1960s. At this time there was a general call for geography to become more relevant, to help tackle and solve pressing social problems. This was prompted by militant reactions to issues such as the Vietnam War, urban poverty, racial inequality and increasing levels of debt among developing nations. It was felt that the quantitative, positivist geography had failed to address these problems. Quantitative geography was accused of naively ignoring the inherent consequences of the capitalist system particularly the production of inequality. Neo-Marxist geography developed out of a critique of this system (Cloke *et al.* 1991).

Neo-Marxist urban scholarship has not formed a neat, coherent body of work despite deriving from a common ideological position (Short 1984: 3). Neo-Marxist urban geography has displayed both disputes between different authors and abrupt breaks within the work of single authors as

neo-
Marxist
urban
geography

they have whole-heartedly abandoned previous positions. The two most influential figures within neo-Marxist urban geography have been Manuel Castells and David Harvey and their works demonstrate these tensions well (Bassett and Short 1989: 181).

Manuel Castells's two most influential books were *The Urban Question: A Marxist Approach* (translated into English in 1977) and *The City and the Grassroots* (1983). Both of these books were concerned with the relations between economic and social structures and spatial structures. *The Urban Question* provided a very abstract and theoretical reading of these relations, something for which Castells was frequently condemned by his critics (Bassett and Short 1989: 183). He was particularly concerned with the role of the state as a regulator of urban crises. These crises, he argued, following a well-worn Marxist tradition, derived from the contradictions inherent in the capitalist mode of production. *The City and the Grassroots*, as the name suggests, was a subsequent attempt to include human agency within his Marxist framework. This he attempted through a number of case studies of urban social protest movements and their influence within urban change. *The City and the Grassroots* was a recognition that dominant class ideologies and the imperatives of economic relations were not unproblematically stamped across space. Rather, spatial relations reflected the patterns of resistance and opposition that these imperatives met. For a more complete analysis of these relations between economics, society and space it was important that this resistance was recognised (see Bassett and Short 1989: 181–3).

The early work of David Harvey, for example *Social Justice and the City* (1973), represented an attempt to read historical cycles of urban development as a reflection of the resolution of crises of over-accumulation within various 'circuits of capital'. This is an approach that attempts to link urban restructuring to wider processes of economic restructuring. It focuses on the built environment as a destination for investment, the profitability of which is linked to the state of the wider economy. Harvey argued that investment in, and hence production of, the built environment occurred when an over-accumulation of capital in manufacturing and commodity production caused returns in this sector to fall. This made land and property an attractive alternative investment. Providing that a framework existed to facilitate it, these conditions caused capital to 'switch' from the former to the latter. Marxist terminology referred to capital switching from the 'primary' to the 'secondary' circuit. However, this capital switching and the patterns of urban growth tend to take on a cyclical form. Investment in the secondary circuit will tend to lead eventually to an over-accumulation in this sector, causing returns on investment to fall. The result of this is that capital will either switch back to the primary sector or seek more profitable investment opportunities within the secondary circuit. These are to be found in newer developments. This switching of capital is manifest in the built environment becoming abandoned in the wake of capital movement to more profitable opportunities elsewhere. Marxist economic theory applied by Harvey saw the built environment as the site for the temporary and somewhat unstable resolution of crisis of over accumulation in the capitalist city (Savage and Warde 1993: 46–7). The examples that have been used to support Harvey's observations include the growth of post-war US suburbs and the office boom in the UK in the 1970s and 1980s.

However, this approach does have its limitations. Savage and Warde (1993: 48–50) in a review of Harvey's contribution to urban geography outline a number of these. Harvey's account of both the built environment and social struggle is partial. It omits a number of crucial dimensions. Harvey suggests that capital switching within the secondary circuit involves a change of location. This is not necessarily the case. It ignores the conversion of property, factories to shopping arcades for example. However, just because capital may not actually switch location does not mean that these changes will not have social consequences. The workers made redundant from factories are unlikely to be those who get jobs in luxury shopping arcades and wine bars. Even if they did, the rewards would be unlikely to match those they received in their previous employment. Harvey's account of social struggle relies heavily on a conceptualisation of this struggle organised along the lines of class. While not necessarily theoretically flawed this is at least limited. Harvey does not take much account of groups based on lines other than class. There are many cases of groups based on gender, ethnicity and sexuality having significant influences on the process of urban restructuring. For example, Manchester in the UK and New York, Minneapolis, San Francisco and other cities in the USA have had gay groups who have been influential in the development of their inner areas (Lauria and Knopp 1985; Knopp 1987). Such alliances may cut across class lines. Perhaps the most serious limitation of this approach is that it has failed to sustain any significant or wide-ranging research programmes. Harvey's own work is light on detailed examples to back up his observations and few others have stepped in to fill this vacuum (Savage and Warde 1993: 48). While Harvey's ideas have generated plenty of debate in academic circles, much of this has been very abstract and little has been rooted in concrete explorations of examples of urban restructuring. For example, theoretically some questions remain about Harvey's failure to distinguish the causes and the consequences of capital switching. While these might normally be expected to be ironed out in subsequent applications of theoretical ideas this has not been the case with Harvey's work (Bassett and Short 1989: 183–6; Savage and Warde 1993: 45–50).

Urban sociology

The relationship between urban geography and urban sociology has traditionally been close. The interchange of ideas dates back to the 1920s with the production of Burgess's model of concentric zones (Figure 1.1). This was the product of the work of the Chicago School of urban sociology that later formed a bedrock of research and teaching in urban geography. Urban sociology, like urban geography, has been far from static and has passed through a number of theoretical developments and debates. Urban sociology has been particularly influential on the practice of urban social geography. Some of the most influential work in urban sociology has stemmed from a body of work referred to as *neo-Weberian*, reflecting the influence of the sociologist Max Weber. This has offered a perspective on the city as a site of the regulation and allocation of scarce resources. Pioneering work in this vein was carried out by John Rex and Robert Moore, who investigated the concept of 'housing classes' through research into ethnic minorities and their access to housing in inner Birmingham. This resulted in the classic studies *Race, Community and Conflict: A Study of Sparkbrook* (Rex and Moore 1967) and *Colonial Immigrants in a British City* (Rex and Tomlinson 1979). In these studies they argued that people's access to housing was not simply dependent upon their job, but a host of other factors, including ethnicity (Savage and Warde 1993: 68). It was argued, for example, that 'from the point of view of housing amenities, it is better to be a white labourer than an Asian chartered accountant' (D. J. Smith 1977: 233). Research uncovered examples of racism by private landlords, local authorities and building societies to substantiate these claims. Although in itself a diverse body of work that has evolved a great deal from its origins in the 1960s, it has highlighted the importance of the role of public and private 'gatekeepers' such as the local authority, estate agents and building societies, and the role of consumption rather than production in the creation and maintenance of social divisions (for a summary see Ley 1983: 280–323).

The neo-Weberian approach has fallen out of favour as sociologists have found the ideas of social classes derived from both housing and employment situations to be too broad to be usefully used as an analytical tool because they conflate the essentially distinct areas of employment and housing (Saunders 1990; Savage and Warde 1993: 69).

Urban theory in the 1990s

Urban theory in the 1990s is in a state of some uncertainty. No one philosophical perspective holds ascendancy. Indeed one of the few positions that unites urban geographers is their wariness of the grand claims of totalising urban theories. One negative consequence of this has been the tendency among urban geographers to shy away from overt theoretical debate. However, more positively this lack of any single philosophical hegemony has opened up urban geography to the application of an eclectic range of perspectives. 'Readings' of the city are more likely to encompass perspectives derived from literary theory, film studies, psychoanalysis and cultural studies as they are Marxist economics or neo-Weberian sociology.

This eclecticism and cynicism of grand theory are a result of the feeling that the urban theories outlined in this chapter are unable to provide anything more than partial accounts of the city. Further, they are becoming increasingly remote from the new forces affecting the development of cities, for example, new technologies, new forms of governance, new economic forces and new ecological concerns. It is unlikely, however, that totalising urban theories will ever vanish entirely. Indeed, it is to be hoped that they do not. Much of the intellectual development of urban geography has derived from the debates opened up by its clashing philosophical adherents. However, this is an appropriate point at which to argue that for urban theory to offer a useful contribution it must do two things. First, it must grow and change as cities develop. Theoretically informed accounts of the 'electronic' city and the 'sustainable' city are as imperative in the late twentieth century as they were of the industrial city in the mid-twentieth century. Second, urban theory must be rooted around real urban issues. There are a range of continuing and very pressing urban problems affecting the world's metropolises. These will not go away as cities change. Contrasting perspectives on these rather than wildly abstract arguments can only enhance the subject's relevance.

This book does not attempt to develop any new or alternative totalising urban theory. Indeed much of it is written wary of any such theory. Rather, it aims to offer a series of syntheses of significant bodies of literature on the emergent forces of urbanisation and the urban changes of the late twentieth and early twenty-first centuries.

Further reading

Excellent discussions of the development of urban theory in geography and sociology can be found in:

Bassett, K. and Short, J. (1989) 'Development and diversity in urban geography' in Gregory, D. and Walford, R. (eds) *Horizons in Human Geography* London: Macmillan, 175–93.

Savage, M. and Warde, A. (1993) *Urban Sociology, Capitalism and Modernity* London: Macmillan (Chapter 2).

A series of concise and useful summaries of key ideas can be found in:

Johnston, R.J., Gregory, D. and Smith, D.M. (eds) (1994) *The Dictionary of Human Geography* 3rd edn Oxford: Blackwell.

Good introductions to the philosophical developments of geography in general are:

Cloke, P., Philo, C. and Sadler, D. (1991) *Approaching Human Geography: An Introduction to Contemporary Theoretical Debates* London: Paul Chapman.

Johnston, R.J. (1983) *Geography and Geographers: Anglo-American Human Geography since 1945* London: Edward Arnold.

Johnston, R.J. (1983) *Philosophy and Human Geography* London: Edward Arnold.

A recent consideration of the possibilities of urban theory is:

Cooke, P. (1990) 'Modern urban theory in question' *Transactions of the Institute of British Geographers* (ns) 15, 3: 331–43.

If you want to explore the history of urban theory further you could sample some of the classics from various schools; a positivist classic is:

Harvey, D. (1969) *Explanation in Geography* London: Edward Arnold.

A classic humanist critique of the modern urban environment and the forces that have created it is:

Relph, E. (1976) *Place and Placelessness* London: Pion.

Castells's Marxist accounts mentioned were:

Castells, M. (1977) *The Urban Question: A Marxist Approach* London: Edward Arnold.

Castells, M. (1983) *The City and the Grassroots* London: Edward Arnold.

A broad cross-section of David Harvey's writings can be found in:

Harvey, D. (1989) *The Urban Experience* Oxford: Blackwell.

Good examples of the neo-Weberian approach in urban sociology are the works by John Rex, Robert Moore and Sally Tomlinson. Many a dog-eared copy can be found lurking in the darkest corners of most university and college libraries.

Rex, J. and Moore, R. (1967) *Race, Community and Conflict: A Study of Sparkbrook* Harmondsworth: Penguin.

Rex, J. and Tomlinson, S. (1979) *Colonial Immigrants in a British City* London: Routledge and Kegan Paul.

③ Economic problems and the city

- The world economy – urban impacts
- The deindustrialisation of the city – geographical consequences
- Theories of economic change

Cities and the world economy

It has become clear that urban development is fundamentally influenced by position in the world economy. This raises important questions about how we understand this process. First, it suggests we cannot understand the processes that shape and reshape cities by only looking within cities. We must adopt a much wider perspective, one that recognises that cities are shaped by processes from far beyond their boundaries, as well as factors much closer to home. Despite this we must not lose sight of the fact that cities are not the helpless pawns of these processes. These global forces are mediated locally. Namely, their outcomes are determined by local factors such as the nature of local urban governments, economies and cultures. Cities are shaped by the interplay of local, regional, national and international forces (Healey and Ilbery 1990: 3–6).

Second, while the world economy is becoming increasingly inter-connected, through the international operations of multinational or transnational corporations, the international dimension of urbanisation is not new. Cities have long performed international functions and many have been profoundly shaped by these. Older industrial cities of the UK have long traded with countries all around the world. London was the command centre of a world-wide empire. The legacies of these international functions are still apparent in the landscapes, economies and institutions of cities as well as the links between cities.

Despite long histories of international functions there are a number of reasons why the recognition of the international contexts of urbanisation is of renewed significance in the 1990s. Previous accounts of urbanisation have tended to under-recognise the significance of the international dimension of urbanisation. They must be regarded, therefore, as partial accounts. This came at a time when the adverse effects of international competition were being felt in the economies of many of the older industrial cities of Europe and North America, international organisations such as the European Union were becoming increasingly significant shapers of fortunes locally and an increasing number of influential companies were operating internationally (Knox and Agnew 1994; Hamnett 1995).

This chapter and the one that follows it consider the international dimension of urbanisation through four questions:

● What have been the main trends in the world economy since the early 1970s?
● What have been the impacts of these trends on the cities of Europe and North America? This will consider the impacts upon the internal structure and operation of cities and upon the relations between cities.
● How have these impacts varied between cities of different types?
● How have cities responded to these changes?

Deindustrialisation and the city

Many of the major cities of North America and Europe were founded, or became closely associated, with the expansion of industrial capital from 1850 onwards. The rise of industry within cities represents a major phase in the history of the city. However, by the early 1980s the majority of these cities were experiencing severe problems with their economies. Unemployment, through the decline of the manufacturing sector, emerged as the major problem facing older industrial cities in North America and Europe. The rapidity and extent of this problem were startling (Table 3.1).

Table 3.1 *Shifts in the relative importance of manufacturing and services in the UK*

| | No. of employees (thousands) | |
	Manufacturing	Services
1975	7 351	12 545
1980	6 801	13 384
1985	5 254	13 769
1990	4 994	15 609
1995	3 918	16 236
Change	– 3 433	+ 3 691

Source: *Employment Gazette / Labour Market Trends*

Looking inside the stark figures revealed that the problem of manufacturing decline in urban areas displayed a number of important dimensions:

Temporal The emergence of a problem of long-term unemployment. Significant numbers of unemployed people remained out of work for periods of well over one year.

Sectoral Unemployment was concentrated in manufacturing which was once a dominant sector of national economies.

Regional Important inter-regional dimensions emerged. Regions such as the North of England and the manufacturing belt of the American Midwest emerged as regions with severe economic problems (see Case Study A).

Urban Cities, the focus of manufacturing industry, bore the brunt of manufacturing decline. This was largely concentrated in their inner areas.

Social The worst impacts of unemployment were concentrated in a number of social groups including youths, the late middle aged, males and ethnic minorities.

The deindustrialisation of the manufacturing cities of North America and Europe can be attributed to three factors: factory closure, the migration of jobs to other areas of the country or abroad, and the replacement of jobs by technology. Each of these will be considered in turn.

Factory closure

Factory closure is a process that has become synonymous with the economies of the inner areas of the large industrial conurbations of the UK in the post-war period. Some of the worst affected cities include Glasgow, Newcastle, Liverpool, Manchester, Sheffield, Birmingham and London. The loss of industrial capacity from these inner-cities dates back to the late 1940s with the beginnings of industrial decentralisation. The process deepened in the period after 1960 and began to include company bankruptcy as well as decentralisation. The prospects for manufacturing were not helped by massive disinvestment in the industrial capacity of the inner-city from the 1960s onwards. Between 1960 and 1982 every major industrial conurbation in the UK lost between one-quarter and one-half

Case Study A

Long-term unemployment on Merseyside

The development of once prosperous Merseyside was predominantly based on its dock facilities. While a boon in economically buoyant times this concentration on a single sector can become a burden in times of recession, creating a spiral of decline that can be broken only through the diversification of the economic base. Merseyside's failure to do so is reflected in the extent and nature of its current unemployment problem. Merseyside lost 199 000 jobs between 1977 and 1990, total employment falling from 619 000 to 420 000. Its 1993 unemployment rate was 17.5 per cent compared to a national average of nearer 10 per cent. The majority of the jobs lost were manual jobs.

The most chronic aspect of the unemploy-ment problem was the emergence of a group of 47 000 long-term unemployed, the majority male, manual workers. The prospects for this group are particularly severe. The majority of new jobs in Merseyside are in the service sector which requires an educated, skilled, flexible labour force, characteristics that are rarely found in long-term unemployed people. The long-term unemployed are also geographically immobile, half of them living in council housing.

Ironically, should Merseyside successfully diversify its economy this group may find themselves increasingly economically marginalised. Their skills, or lack of them, are appropriate only to routine production in the manufacturing sector. However, leisure and office developments, at the forefront of urban regeneration schemes across the UK, are likely to increase land values above what manufacturing can afford, precluding the reindustrialisation of the region.

Source: Hamilton-Fazy (1993)

of its employment. By far the worst affected areas were the inner-cities. The most rapid of this decline occurred in the post-1970 period. The rise in unemployment among the skilled, male workforce of the UK's inner-cities was largely a result of factory closure in these areas not being compensated by new companies opening (Massey 1988; Champion and Townsend 1990).

Factories in inner-city areas closed for one of three reasons: company bankruptcy, companies switching production and companies switching location (Hudson and Williams 1986: 107). The bankruptcy of companies can be largely attributed to the rise of cheaper manufactured products from abroad coupled with products and methods of production that were reaching the end of their life cycles that were not replaced and restrictive government policies. This was due to a lack of innovation and investment within British industry.

Since the early 1980s the actual rates of factory closure have declined in the UK although the general retrenchment of the industrial sector has continued (Healey and Ilbery 1990: 321). Some reindustrialisation of the inner-city has been attempted through the development of science, business and industrial parks and policies such as Enterprise Zones and Urban Development Corporations (see Chapter 5). However, the successes of these have been limited when compared to the problems created by deindustrialisation and they have brought with them their own problems.

Migration of jobs to suburban and rural locations

This migration was the combination of the emergence of problems with inner-city locations and advantages of suburban and rural locations. It reflects changes in the nature and requirements of manufacturing industry and in the organisation of the national transport infrastructure. The suburbanisation of industrial production was primarily related to problems with inner-city sites relative to new suburban locations. These problems included inner-city congestion and a lack of space to accommodate site expansion and the development of assembly line production techniques. By contrast the suburbs offered cheaper, extensive sites accessible by the growing motorway network. Location switches were also facilitated by the fact that much industrial capital in inner-city sites was old and in poor condition. It was economically more feasible to write this off and relocate with new machinery than undergo in situ upgrading of capital. The office boom of the 1960s and 1970s had also forced up urban land prices. This made selling some inner-city factory sites for redevelopment profitable, thus further facilitating moves to the suburbs (Hudson and Williams 1986; Massey 1988; Champion and Townsend 1990).

The spatial dimension of economic change is not independent of other dimensions. The suburbanisation of industry has important implications for the gender composition of the industrial labour force and for social relations within the home. Industrial employment in the inner-city was traditionally, although not exclusively, male. However, suburban industrial labour employs a high proportion of women. The lack of a developed culture of unionisation and militancy and a tradition of skilled,

highly paid work among this group have made them attractive to
employers who have hired them at cheaper rates and demanded greater
flexibility from them than was the case with 'traditional' male industrial
labour forces. Clearly this shift has transformed the economic relations
within many households as women have found themselves increasingly
the main 'breadwinner'. The transition has proved problematic in many
areas where the culture of male employment is deeply rooted (McDowell
and Massey 1984).

Migration of jobs internationally

The decentralisation of production to newly industrialised countries or
countries previously peripheral to the world economy has attracted a
great deal of attention in the context of the deindustrialisation of older
industrial cities in the UK, Europe and North America (see Case Study
B). The main agents responsible for this decentralisation of production
have been those large companies known as transnational, or multinational, corporations (Dicken 1986; Knox and Agnew 1994). These
companies have emerged as the key players in the world economy since
the mid-1970s. Multinational corporations transact a significant
proportion of their business in one form or another, outside their home
economy. They are not dependent for their welfare on the state of their
home economy, or any other single national economy for that matter
(Allen 1995: 6). Although multinational corporations are not new, their
independence from national economies and the sheer extent to which
they influence the world economy is.

The multinational corporation offers an organisational size and structure
with a number of advantages in the current international environment.
They are able to use spatial differentiation, most notably in labour costs,
to their advantage. Locating production units in the metropolitan areas of
newly industrialised countries allows them to tap the pools of cheap,
unskilled, non-unionised labour there (Dicken 1986; Allen 1995: 85–8).
This decentralisation has been facilitated by innovations in production
technology. These have lead to a decline in the skill levels required in the
production process and have allowed the spatial separation of different
tasks in the production process which in turn has opened up the
possibility of redistributing production units to take advantage of labour
cost differentials across space (Dicken 1986: 204). This spatial
fragmentation has been 'glued' together by developments in

telecommunications technology (Sassen 1994; Graham and Marvin 1996). These locations bring other advantages such as incentives from national governments and primitive or non-existent worker protection legislation. The multinational corporation also offers an organisational form that is flexible. Multinational corporations typically employ a range of flexible organisational arrangements such as joint-ventures and subcontracting production out to other firms. Multinational corporations also have the capacity to accommodate extensive research and development infrastructures. These allow constant innovation such as product modification and development (Dicken 1986; Knox and Agnew 1994; Allen 1995).

Automation of production

The main adverse impact of technology was the substitution of technology for human capital. Whereas in the period up to the late 1960s technological innovations created new products, stimulated new demand and consequently increased manufacturing employment, the reverse became true during the 1970s. Automated production began to replace routine, unskilled and semiskilled tasks in older manufacturing cities (Dicken 1986: 400). As with most of these factors the impacts of increased automation were concentrated sectorally, socially and geographically.

Consequences of deindustrialisation

Not all cities in North America and Europe were equally affected by the processes of deindustrialisation. Those cities with diverse economies or without a significant manufacturing component in their economies enjoyed very different economic fortunes during the period of deindustrialisation. It is also difficult to generalise about the impact on different industrial cities. The impacts of deindustrialisation varied between different industries and depended upon the particular composition of individual urban economies and national and local government actions. However, despite these qualifications it remains true to say that deindustrialisation was the most significant economic process to affect large cities in Europe and North America since the 1960s.

One of the most dramatic demonstrations of the loss of economic dynamism in the urban industrial sector has been the fall in the total

Case Study B

Alternative theories of economic change

One of the major problems outlined in the previous discussion was the deindustrialisation of former manufacturing cities. While much of this was due to the relocation of industry to suburban and rural locations a significant proportion was the result of relocation of assembly plant to newly industrialised countries or closure due to import penetration from these countries. It has been argued that the shift of industrial capacity to newly industrialised countries represents the emergence of a new international division of labour in which the manufacturing functions of the inner areas of older industrial cities has been surpassed.

This theory is able to account for a number of major, world-wide economic developments. These include the deindustrialisation of Western cities, the growth of cities in newly industrialised countries and the growth of global cities as the control and command centres of an interconnected world economy. However, despite this, the explanatory scope of this theory, while not being incorrect, is limited. In relying so heavily on economic processes it is able to say little about, for example, the social geographies of cities which are clearly related to economic change. This theory is able to offer only one-directional explanations of the relationship between economic change and urbanisation. This perspective is able to demonstrate how macro-economic change has impacts on cities. However, it can say little about how the characteristics of individual cities, their political leadership or local business community for example, can affect the operation of macro-economic processes. It is able to look at the city from the outside-in, but it fails to be able to see from the inside-out. Cities do not find their position in the international division of labour automatically determined by their geographical location or their economic histories. They may be constrained by these factors but they are able to influence the operation of the economic processes affecting them. The new international division of labour perspective has little to say on these important aspects of the relationship between economic change and urbanisation.

An alternative theory that adopts a far less abstract approach than the previous one is the 'restructuring' theory. The 'restructuring' refers to the restructuring of organisations in response to changing economic conditions. This approach was primarily developed by two geographers, Doreen Massey and Richard Meagan, in the late 1970s and was refined during the early 1980s in publications such as *The Anatomy of Job Loss: The How and the Why and the Where of Employment Decline* (Massey and Meagan 1982) and *Spatial Divisions of Labour* (Massey 1984). Massey and Meagan noted that spatial unevenness in rates of employment and unemployment had come to characterise the UK economy. This, they argued, was the result of the spatial restructuring decisions undertaken by companies and organisations. Companies were able to use space – differences in labour costs across space, for example – to their advantage. Restructuring was organised around exploiting these spatial variations in an attempt to maintain profits in the face of an increasingly competitive world economy. Patterns of uneven national and international development reflected the ability of organisations to use spatial variations, not only in labour costs, but also in the socio-cultural characteristics of areas (history of unionisation and militancy, for example).

The deindustrialisation of the inner areas of older industrial cities can be interpreted as a restructuring response by manufacturing companies. These areas were characterised by high labour costs, and militancy compared to suburban and rural locations. Massey and Meagan described how regional specialisms developed as the production process became broken down into its component parts and became spatially dispersed. This was able to explain the emergence of the South East region of the UK as an area with a high concentration of headquarters and command and control functions while older industrial regions became the sites of routine production in plants controlled and owned elsewhere. Massey and Meagan looked at the increasing mobility of capital and the increased salience of the social rather than the natural characteristics of areas.

This approach generated a number of positive advances. It sustained a major research programme into the changing urban and regional structure of the UK and it was able to go beyond economistic explanations to show how social and cultural characteristics are implicated in economic restructuring. A major insight to stem from this was the demonstration that the relationship between economic restructuring and the specific geographical make-up of places was a two-way rather than a one-way relationship. Not only did economic restructuring impact on the geographies of individual places but also these unique geographies themselves impacted on the operation of the processes of economic restructuring. The restructuring approach showed how places were not simply the passive receptors of economic change layered down from above but they were active in affecting the outcomes of these changes.

Despite this the restructuring approach was the subject of increasing criticism during the late 1980s and 1990s. In contrast to other approaches the restructuring approach poorly articulated the impacts of economic restructuring upon specific places. Research in the restructuring vein has tended to suggest that places are far more coherent and heterogeneous than they actually are. The 'localities', as they were termed, in which this research was conducted were all relatively small and represented quite distinct areas. However, even these demonstrated a great deal of internal diversity, spatially, economically and socially. The impacts of economic restructuring on these localities was,

Case
Study B
continued

therefore, dispersed throughout the social and economic structure of the locality. The impact depended very much on one's location within this structure. It is not necessarily meaningful to adopt a spatial unit which suggests some degree of internal homogeneity to study these impacts. Overall the restructuring approach tended to overemphasise the role of space in the mediation of economic restructuring and under emphasise the importance of the internal structure of the locality. The restructuring approach must be regarded as an ambitious but limited attempt to show the importance of social and cultural factors in the mediation of economic restructuring.

Source: Savage and Warde (1993: 41–61)

urban populations of large British cities. This was predominantly due to out-migration, a process known as counter-urbanisation or the suburbanisation of populations beyond municipal boundaries. This was a particularly noticeable characteristic in the 1970s and early 1980s. Cities did show some signs of recovering population in the late 1980s (Table 3.2).

This process displayed a pronounced social dimension with mainly middle-class residents moving to the suburbs or beyond, while more disadvantaged groups were far less mobile. Coupled with the migration of industry out of large urban areas and factory closure, this led to something of an economic vacuum developing in inner-city areas and a spatial polarisation of urban populations based around income, lifestyle and opportunities. Increasingly, the inner-city found itself disconnected from the dynamics of the formal economy and developed, or failed, as a place and a people left behind, as these dynamics shifted elsewhere.

Table 3.2   *Population change in selected European cities 1970–90*

	1970–75		1975–80		1980–85		1985–90	
	City	*Reg.*	*City*	*Reg.*	*City*	*Reg.*	*City*	*Reg.*
London	−1.89	−0.37	−1.60	−0.14	−0.38	−0.06	0.56	−0.32
Birmingham	−0.3	0.35	−1.01	−0.66	−0.33	0.00	−0.37	0.06
Glasgow	−3.38	−1.47	−1.84	−0.11	−1.06	−0.17	−1.44	−0.32
Hamburg	−0.77	0.85	−0.91	0.36	−0.77	0.06	0.24	0.06
Paris	−1.48	1.93	−0.69	0.66	−1.02	0.78	1.01	2.06
Amsterdam	−1.84	1.51	−1.11	0.81	−1.18	0.57	0.34	0.47

Source: Based on *A Report to the Commission of the European Communities Directorate General for Regional Policy* (XVI), April 1992 p. 56, cited in Sassen (1994: 41)

Future reading

The accounts given here of the major economic changes that have affected the city are by necessity very brief. More detailed accounts can be found in:

Dicken, P. (1986) *Global Shift: Industrial Change in a Turbulent World* London: Harper and Row.

Healey, M. and Ilbery, B. (1990) *Location and Change: Perspectives on Economic Geography* Oxford: Oxford University Press.

Knox, P.L. and Agnew, J. (1994) *The Geography of the World Economy* 2nd edn London: Edward Arnold.

More detailed discussions of the transformation of the UK economy can be found in:

Allen, J. and Massey, D. (eds) (1988) *The Economy in Question* London: Sage.

Champion, A. G. and Townsend, A.R. (1990) *Contemporary Britain: A Geographical Perspective* London: Edward Arnold.

Hudson, R. and Williams, A. (1986) *The United Kingdom* London: Harper and Row.

Detailed accounts of different theories of urban development can be found in:

Savage, M. and Warde, A. (1993) *Urban Sociology, Capitalism and Modernity* London: Macmillan / British Sociological Association.

4 New urban economies

- Cities and the service economy
- New economic geographies of cities
- Telecommunications and cities
- The 'urban doughnut'

Cities and the rise of the service economy

A great deal of optimism has been placed on the rise in service sector employment since the early 1980s. It was felt that the rise of this alternative sector might offset the losses experienced in the industrial sector in many cities of Europe, North America and Australia. However, the rise of services was sectorally, socially and geographically specific. Not only were they unable to fully compensate for loss of manufacturing jobs but also the places and people who benefited from them were very different from those who bore the brunt of manufacturing deindustrialisation (Hudson and Williams 1986: 112; Allen 1988).

Service sector employment rose for a number of reasons in the early 1980s, these included the demands of businesses for specialised financial and legal services, the co-ordination required to orchestrate spatially dispersed economic activities within companies and the increased demands of households for services. Between 1945 and 1990 service sector employment grew by approximately 75 per cent. This growth was sectorally uneven with particularly rapid growth in the distributive and banking and insurance producer services (Cameron 1980; Allen 1988; see Table 4.1).

The absolute rise in the number of service sector jobs has been significantly less than the absolute number of jobs lost in the manufacturing sector. This is reflected in rises in the total number of

Table 4.1 *Rapid growth of various service subsectors in the UK*

	No. of employees (thousands)			
	Distribution, hotels and restaurants	*Banking, Finance, insurance, etc.*	*Education*	*Medical, health and veterinary services*
1975	3 906	1 468	1 534	1 112
1980	4 206	1 669	1 586	1 214
1985	4 213	2 039	1 557	1 301
1990	4 756	2 701	1 735	1 450
Change	+ 850	+ 1 233	+ 201	+ 338

Source: *Employment Gazette*

unemployed people. This is clearly demonstrated by the case of the UK economy which, despite a significant rise in service sector employment, suffered a rise in the total unemployed between 1981 and 1991. What these absolute figures fail to demonstrate, however, are the resultant social and spatial dimensions of the change in the composition of employment.

The regional and urban impacts of the rise of the service sector and the broad shift from manufacturing to services within the UK economy, for example, can be interpreted as a complex interplay between social, economic, temporal and sexual dimensions. The broad impacts were an increase in the overall rate of unemployment, the transformation of local labour markets and the spatial decentralisation of the service sector across the UK.

First, the rise in service sector employment has been unable to compensate fully for the loss of manufacturing jobs. This broad shift displays a markedly uneven regional dimension. The worst affected areas in terms of overall increases in unemployment have been the urban areas of former manufacturing 'heartlands', for example, in the Midlands (Spencer *et al.* 1986), the North of England and the Celtic 'fringe'. This has included areas such as South Wales, the West Midlands, the North East, Clydeside in Scotland and Northern Ireland (Champion and Townsend 1990). Similar patterns have emerged in both North America and Australia. A distinctive rust-belt of north-eastern cities has emerged in which heavy manufacturing job loss has occurred. These cities include Baltimore, Pittsburgh, Cleveland, Detroit and Chicago. Economic growth in the USA has been largely spatially distinct from these areas of decline, focusing on an emergent sun-belt in, for example, California and involving industries utilising new technologies such as animation, biotechnology, and space research (Knox and Agnew 1994: 247). Australia's rust-belt has included the states of Victoria, Tasmania and South Australia, while a sun-belt has emerged including Queensland and Western Australia (Stimson 1995; see also Case Study D, pp. 52–3). Sectoral economic shifts inevitably involve transformations in the 'space economy'. The use of the climatic metaphor sun-belt is appropriate to describe and explain the emergence of new sectors and areas in the space economy. Often the possession of a pleasant climate is one of the few natural advantages of significance in explaining the growth of new industries. The climate of, for example, California is one of the reasons that the highly qualified employees of new industries are attracted there.

Not only have rust-belt areas been quantitatively worst affected with the most severe retrenchment of male employment coinciding with sluggish service sector growth, but also they have been qualitatively transformed. The new service sector jobs have tended to differ significantly from the manufacturing jobs lost. The impacts of this on labour markets have included the rise of part-time working and flexible work practices such as temporary contracts and 'hire and fire' recruitment, and the increasing involvement of women in the labour force. Some implications of these changes have included the polarisation of income opportunities with the erosion of the intermediate income layer within labour markets and changing social relations within the home as women have progressively replaced men as the family breadwinner in many areas.

The geography of service sector growth is similarly complex. However, in the UK a major aspect of this can be summarised as the selective decentralisation of office employment away from the South East. The office sector in the UK developed rapidly in the 1960s and was heavily concentrated in London and the surrounding South East region. The office concentrations included the headquarters of national and multinational corporations seeking proximity to the financial district of the City of London, the headquarters of newly nationalised industries and the expanding civil service. The early decentralisation of offices was largely prompted by central government action. This included the Office Development Permit that encouraged movement beyond the South East, the Location of Offices Bureau (1964–79) which promoted alternative office locations and the movement of government offices away from

London, for example the vehicle licensing office to Swansea (Hudson and Williams 1986: 111). This was a deliberate attempt to even out the development of office employment across the country. Private companies began to decentralise during the 1960s and 1970s. However, they tended to remain within the South East region, for example Eagle Star (insurance) which moved to Cheltenham and IBM (computers) which moved to Bristol. It was rare to find examples of significant relocations beyond this region.

Telecommunications improvements have created the possibility of the 'virtual office' (spatially distant but electronically interlinked networks of offices within companies) and companies have begun to spatially 'disintegrate' their offices according to function (Bleeker 1994; Graham and Marvin 1996: 128). This has led to some relocation of routine, 'back-office' functions and some middle-management functions to peripheral locations in former manufacturing cities and locations on the Celtic fringe. These relocations have taken advantage of cheaper and more flexible labour. Higher-level management office functions have not relocated to anything like the same extent as more routine functions.

The types of jobs created by the increase in service employment have tended to be polarised between managerial jobs which are relatively small in number and more routine (back-office) jobs which are characterised by low levels of pay, low skill requirements, a lack of training and union representation, poor prospects and part-time or temporary contracts. This polarisation of opportunity has failed to replace the middle section of the labour market (well-paid, full-time, semiskilled jobs) that was devastated by the process of deindustrialisation. Socially these opportunities have failed to reincorporate those made redundant through deindustrialisation. The labour markets of western economies have been characterised by a shift from male to female employment in former industrial areas as well as the national economies generally.

The decentralisation that has characterised the rise of the service economy generally was initially fuelled by advances in tele-communications. However, the further advances in telecommunications threaten to undermine the advantages of peripheral areas in Western nations by allowing the further decentralisation of back-office functions abroad. Again the motivation behind this is the advantages associated with an oversupply of cheap labour in these areas (Graham and Marvin

1996: 153). Although this trend has not developed to a significant extent yet, individual examples suggest it might in the future. For example, the catering aspects of British Airways world-wide booking system are now co-ordinated from an office in India (Graham and Marvin 1996: 154).

New economic geographies of the city

This section examines the ways in which the urban economy is becoming reconstituted around new or growing sectors. It considers the implications of this for different types of cities and for the internal geographies of cities (see Case Study C).

Cities and corporate headquarters

Table 4.2 *Global cities and corporate headquarters*

City	Fortune Global Service 500[a] Headquarters	Fortune Global 500[b] Headquarters
Amsterdam	4	1
Atlanta	5	3
Chicago	7	10
Frankfurt	8	3
London	28	35
Los Angeles	7	10
New York	25	12
Osaka	20	21
Paris	28	23
Sydney	4	5
Tokyo	86	83
Washington, DC	6	5

Source: Knox (1995: 238)

Notes: [a] includes banking, financial, savings, insurance, retailing, transport and utilities services
[b] includes all industrial companies

Corporate headquarters have always displayed an urban bias in their location. While this has remained, except for some national corporations and some small multinational corporations which have relocated to suburban locations or smaller urban settlements, locational change has occurred reflecting shifts in the economic geography of national economies. Manufacturing cities and regions have tended to decline as centres for corporate headquarters. This has been most noticeable in North America where the cities of the Midwest have lost headquarters functions. There has been a general shift in the location of corporate headquarters in Europe and North America from manufacturing cities to those more associated with service economies (Knox and Agnew 1994: 251).

Case study C

European cities in the post-industrial international economy

In a reversal of the trends of the 1960s and 1970s, large cities in Europe and North America began to gain population and grow economically in the late 1980s. The slowing of these growth patterns in smaller cities, previous foci of growth, seemed to suggest that the changing nature of the global economy was transforming the urban systems of these continents. A number of trends began to become apparent in the European urban system:

- several transnational, but sub-European systems have emerged
- a small number of large cities have strengthened their positions within a new European urban system
- a few select cities, such as London, Paris, Amsterdam and Zurich, have become connected, and constitute part of a global system of major cities.

These trends have opened up new divisions between cities in Europe. Those cities that have enhanced their positions within these urban systems have tended to be high technology industrial centres, service centres and well connected to high-speed transport and communication corridors. These corridors, based on high-speed air and rail links and telecommunications, have grown up between major international centres or groups of specialised centres with complementary functions. Becoming hubs of these transport and communication networks has tended to instil major growth dynamics in urban economies and infrastructures, as in the case of the former industrial city of Lille in north-east France.

The major cleavage within urban systems is between the cities described above and cities with outdated economic infrastructures based on industry, defence or port facilities. The problem is particularly acute for cities with scarred or polluted environments, a lack of modernisation and/or a physically isolated location. The emergence of international or global urban systems has reconfigured the urban systems of individual nations. Again the trend is towards polarisation, those cities connected internationally capturing new functions and investment and increasing in importance within their national urban systems, in some cases eclipsing national capitals, those disengaged from international urban systems becoming increasingly peripheralised and removed from major economic growth dynamics. Within the latter group there tends to be little chance of re-invigorating previously important economic sectors or being able to engage with emergent international urban systems. The only hope remaining is the development of new functions. This has occurred in a few select cases, for example, in Lille's emergence as a transport interchange centre.

Case
Study C
continued

Two further influences on the European urban system are the democratic reforms of Eastern Europe and mass migration from outside the European Union. The re-emergence of links between Eastern and Western Europe in the 1990s is likely to favour both those cities in border regions and those Western European cities with historical links to Eastern Europe, such as Hamburg (Germany) and Copenhagen (Denmark). Within Eastern Europe a few major cities are likely to emerge as foci of investment for the region. This has accounted, for example, for the transformation of Budapest's economy and landscape in the 1990s.

The expected increase in immigration into the European Union from Eastern Europe, the Middle East and North Africa is likely to contribute to the development of a number of European cities, both as entrepôts and destinations. Major entrepôts are likely to include Marseilles (France), Palermo and Bari (Italy). Where these cities have suffered deindustrialisation, the extra strain on their economies is likely to lead to further problems. Emerging destination cities include Berlin and Vienna. Immigrants are likely to form a significant proportion of the informal economies and sectors such as construction, in these cities.

The European urban system has been affected by a number of internal and external forces since the early 1980s. These have transformed many of the existing relations between cities, created new ones and destroyed some. The European urban system has developed many complex new sets of centre–periphery relations as a result.

Source: Sassen (1994: 39–47)

The headquarters of large multinational corporations have tended to concentrate in the very largest cities, and have become disproportionately concentrated in a small number of 'world cities' or 'global cities'. London, Tokyo and New York are the most advanced of this numerically small but hugely influential group. Table 4.2 indicates the concentration of corporate headquarters in these cities (Sassen 1991; 1994; Knox 1995).

The consolidation of corporate headquarters in large urban areas reflects their need for access to regional, national and international markets, a highly skilled labour force and a range of sophisticated, specialist service inputs. Only the very largest and well-connected cities are viable to satisfy these requirements. Corporate headquarters in these cities have become the focus of new, dynamic economic quarters that have begun to shape both the central areas and the wider economies of these cities based upon the growth of business (producer) services (Sassen 1991; 1994).

Producer service economies

Producer services offer legal, financial, advertising, consultancy and accountancy services to companies. They have become increasingly important to companies who are required to respond rapidly to changing market conditions which require that they draw upon a range of specialised inputs. These services have become the fastest growing sectors of national economies and the economies of large cities in Europe and North America with significant numbers of international linkages (Thrift 1987; Sassen 1991; 1994; Hamnett 1995; Knox 1995).

Producer services have been overwhelmingly concentrated in the central areas of major world financial centres. They draw on the highly innovative environments of these cities and the extensive and instantaneous links to other industries and experts they offer. As well as the physical and electronic infrastructures of these sites, the social environments of bars, restaurants and health clubs are particularly important to the formation of social networks which are heavily drawn upon in business. This social milieu is impossible to reproduce outside these tight agglomerations of related activities even with sophisticated telecommunications links (Sassen 1994; Graham and Marvin 1996).

The main customers for producer services are the headquarters of multinational corporations. As the increasing extent and complexity of these corporations have proceeded an effective central co-ordination, control and command complex has become ever more crucial. This effectiveness has come to depend to a large extent on the utilisation of specialised producer services. Producer services have become a vital part of the co-ordination of an internationally dispersed but increasingly interlinked global economy. The concentration of corporate headquarters in global cities has provided them with a lucrative, spatially circumscribed market.

Producer services have considerable impacts on the wider urban economies of which they are part. Producer services, along with certain financial and investment services, enjoy a privileged position in their urban economies because of their ability to generate 'superprofits'. Superprofits occur when disproportionately high returns are generated relative to investments of time and money. Because of this, producer services are able to dominate competition for land, resources and investment at the centres of large cities. As other sectors are squeezed out of both the spaces and the lucrative economies of these city centres they become devalued and marginalised. This has a significant social impact because the lucrative opportunities afforded by the producer service sector tends to be limited to a few highly qualified professionals. The remaining jobs in this sector are low-paid, low-skill jobs such as cleaning and security services. This polarisation of opportunity contributes nothing to the alleviation of the disadvantaged position of many inner-city residents. Indeed the impacts may even be negative as locally oriented services are displaced by those such as boutiques and restaurants aimed at producer service workers (Sassen 1994: 54).

Research and development units

The location of company research and development units is dominated by two requirements: access to highly qualified personnel, and proximity to either corporate headquarters or production units. The former tends to exert the greater pull as research and development has become more central to the operation and development of companies in an unstable economic environment (Malecki 1991; Knox and Agnew 1994: 251–2). These requirements tend to be satisfied in one of three locations: large cities with corporate headquarters, near universities or other research institutions – particularly where a pleasant environment acts as a further attraction to qualified staff – or manufacturing sites. The latter site reflects the need, present in some industries, to integrate closely research and development and production. As production has decentralised there is some evidence of associated research and development decentralisation (Knox and Agnew 1994: 252). Should this continue on a significant scale it is likely to rewrite the geography of 'branch-plant' industrialisation, expanding the range and type of linkages between branch-plants and their local economies (*The Economist* 1995). However, it is likely that this decentralisation will be highly selective. Branch-plants in peripheral countries are likely to remain simply as production units, disadvantaged by a lack of indigenous educational and research infrastructure.

The geography of research and development plants is highly significant in urban and regional development. Research and development complexes are sites of innovation where products are modified and new ones developed. They are potentially seed-beds for new patterns of economic growth (Healey and Ilbery 1990: 111–12). A fact recognised by local authorities who eagerly try and encourage the formation of innovative milieux through developments such as science and technology parks, linking universities and business. The problems faced by most of these suggest that it is unlikely that research and development facilities are likely to disperse to any great extent except in association with branch-plants. In the future it is likely that existing centres are likely to consolidate the advantages they already possess.

New industrial spaces

A number of economic activities are beginning to emerge that are mediated through and based upon new technologies including computer hardware and software, telecommunications, virtual reality technology and biotechnology. The nature, organisation and markets of these activities are sufficiently different from those of previous eras for them to be considered new industries. With new industries come new geographies of industrial location sufficiently different from previous forms of industrial locations for us to talk of 'new industrial spaces' (Knox and Agnew 1994: 252–3; Graham and Marvin 1996: 158–9).

These new industries are dominated by three locational requirements: the need for access to a highly qualified and functionally flexible labour force, an environment that facilitates constant innovation and cross-company and industry co-operation and good infrastructural and telecommunication linkages to corporate headquarters, universities and other research institutions, linked companies and national and international markets (Graham and Marvin 1996: 158–9). These industries tend to be attracted by the social, rather than the natural, characteristics of places.

Far from offering hope for the reindustrialisation of problem areas in older industrial cities, the locational and economic characteristics of these industries suggest they are likely to be harbingers of problems as well as possibilities. New industrial spaces are opening up in very different locations to old industrial spaces. The economic regeneration stemming from these developments is unlikely to directly or indirectly benefit those areas which have suffered from deindustrialisation. Emerging trends in industrial and economic geography suggest there is little chance that the dynamic economic mechanisms similar to those associated with industrialisation will return to inner-city locations.

Like other emergent sectors of urban economies, new industries appear to be dominated by a polarised income distribution requiring a small number of highly paid specialists and a large number of low-paid workers. Opportunities extended by this wave of development to those disadvantaged by the abdication of the industrial sector from urban areas are likely to be restricted to the latter. Whereas in the past the 'career-ladder' mechanism offered a pathway for career advance there appears to be no evidence that low-paid and high-paid sectors in these industries are thus linked (Markusen 1983; Knox and Agnew 1994: 253). Working practices in these industries are likely to be dominated by the requirement to be flexible. They are likely to be dominated by part-time working and short-term or temporary contracts. The effects of these new industries on the socio-economic welfare of disadvantaged groups is likely to be negligible, even negative.

Finally, there are question marks over the extent to which these industries are able to generate positive economic linkages to their local areas. These industries generate more linkages within companies and with national and international economies than they do with their local and regional economies (Turok 1993).

Cities, telecommunications and the world economy

The world economy appears to be becoming increasingly interdependent. The fortunes of individual places are not autonomous, they are increasingly bound up with the fortunes of other places and with processes operating at wider geographical scales. This increasing interdependence within the world economy is the result of a number of related processes: the emergence of multinational corporations as major shapers of international economic flows, the concentration of international command centres in global cities, the deregulation of national financial markets and telecommunications advances 'gluing' together spaces and creating the 'space of flows' between cities in the world economy.

The most noticeable effect of this is that certain spaces are moving closer together. What this means is not that their geographical locations are changing but interactions between them are becoming ever more instantaneous and 'real' (Hamnett 1995; Knox 1995; Leyshon 1995). Advances in telecommunications – telephone, fax, e-mail, computer networks, the Internet and virtual reality – have reduced or eliminated the time delay in communication between distant spaces and increased the sophistication of exchange available. However, this 'electronic economy'

Case study D

Australia's cities in a changing national and global economy

The space economy of Australia has been transformed by a set of processes which have been focused upon its major metropolitan economies. Australia's recent economic development has been characterised by the contrasting fortunes of a rust-belt, including the states of Victoria, Tasmania and South Australia, and a sun-belt focused primarily, but not exclusively, on the eastern seaboard cities and new metropolitan regions including those of Queensland, Western Australia and New South Wales. Australia's economy, like that of the UK and USA, has broadly shifted away from manufacturing and towards services. Spatially this has meant that cities with major industrial concentrations, for example Melbourne and Adelaide, have suffered badly in the recession that has affected the country in the 1990s, while cities and emerging urban areas with more diversified or service-oriented economies, such as Brisbane, have fared better.

Australia's industrial troubles date back to the oil price crisis of the 1970s. Its industrial base had developed behind a very protective tariff barrier. The oil price crisis acted as a shock to Australia's industrial sector, necessitating considerable restructuring of its production processes. The world recession that followed created conditions of high inflation and high unemployment within the Australian economy. When the tariff barriers were reluctantly dismantled Australia's industrial sector proved to be exposed and uncompetitive in the face of international competition. Other major changes in the Australian economy at the time included the deregulation of its financial markets and their integration into the global economy. The protected phase of Australia's economic expansion was over and it had to cope with a new set of problems and face up to a new set of challenges. The adjustments were stamped all across Australia's economic landscapes, particularly those of its cities.

A series of positive economic growth sectors emerged in the 1980s which were concentrated spatially within the sun-belt states, particularly their state capitals. The major growth sectors of the Australian economy were producer services (O'Connor and Edgington 1991), media and publishing (O'Connor 1991), construction and tourism. All of these sectors have shown a high tendency to concentrate in Queensland and New South Wales, particularly in the case of producer services in Brisbane and Sydney.

Three further dimensions have affected the geography of Australia's economy. These are the emergence of new industrial spaces focused on research and development and hi-tech manufacturing, a massive increase in foreign direct investment, particularly from Asia, and a refocusing of trading relations towards the Asia-Pacific region. Again the growth dynamics resulting from these trends are focused largely on the sun-belt cities and new urban areas of the country's eastern seaboard. These changes have reflected, and in part accounted for, the shift in Australia's economy away from heavy manufacturing towards information and knowledge based industries and services and property.

Source: Stimson (1995)

is extremely predominantly located in a small number of major cities which contain high concentrations of these technologies and which are highly interconnected with others. Again it is the global cities such as London, New York, Tokyo and Paris which dominate. As one moves down the urban hierarchy, concentrations of advanced communications technologies decline as does the degree of interconnection with the world economy. These patterns show a high degree of primacy. As Table 4.3 shows, the largest cities within individual nations tend to account for a disproportionate percentage of that nation's international communications infrastructure (Sassen 1991; 1994; Knox 1995).

While the high concentrations of international, electronic communications infrastructure mean that a few select places are moving closer together, its converse is that a vast number of spaces with low concentrations of this infrastructure are failing to move closer together. This exclusion distances a large number of older industrial cities, cities in newly developing and less-developed countries and rural areas from the centres of the global economy, which is effectively moving them further apart (Sassen 1994; Leyshon 1995). This is creating a geographical pattern of a small number of 'information-rich' global cities forming a highly interconnected transnational urban system, surrounded by vast 'information poor' hinterlands with which they are poorly connected. It

Table 4.3 _Urban dominance of telecommunication investment and use_

City	Percentage of national population 1994	Measure of percentage of international communications infrastructure 1994
New York	6	35[a]
London	16	30[b]
Paris	18	80[c]
Tokyo	10	37[d]

Source: _Financial Times_ (1994), cited in Graham and Marvin (1996: 133)

Notes: [a] percentage of US outgoing calls starting in New York
[b] percentage of UK's mobile calls made within London
[c] percentage of French telecoms spending
[d] percentage of Japanese telexes in Tokyo

has been argued that as the interconnections between cities within transnational urban networks increase, the connections between these cities and both their regional hinterlands and domestic national urban systems decrease (Sassen 1994). While this is clearly an important emerging pattern, it would be premature to say that these cities are *more* connected to each other than they are to their own regions and nations. Traditional internal linkages between these cities and their regions are being complemented, and in some cases replaced, by flows of money, information, power and people from international cities (Hamnett 1995; Knox 1995). Whether they become usurped by them remains to be seen.

The geography of the international, electronic communications economy displays spatial, urban, sectoral, social and sexual divisions. This economy is overwhelmingly associated with the nations of North America, Europe and Japan. The extent to which cities in newly developing and less developed countries are linked is very limited. Linkages may exist in their capital cities but these are usually of a very limited extent and are generally of a subordinate nature to the international economy, rarely being significant shapers of the geography of international flows. Within North America, Europe and Japan this economy is primarily located in cities of the service economy. Older industrial cities find themselves poorly connected and either are only the receivers of flows rather than the origins and the shapers of these flows, or are peripheral to them entirely. Access to advanced telecommunications tends to be highly restricted socially. They are primarily associated with a small number of professional, managerial jobs. Consequently a large proportion of the population form a massive information-poor underclass. This social dimension has a spatial expression with the inner-city being the most obviously information-poor environment, an 'electronic ghetto'. Finally the majority of jobs associated with the centre of the global information economy are occupied by men. This is, consequently, a very male space.

The 'urban doughnut': the new economic and social geography of the city?

Geographically the pattern of spaces of economic dynamism and economic depression appear to have become increasingly polarised. City centres have suffered a very mixed fate. The centres of a small number of global cities have boomed, fed by the rapid growth and superprofits of the financial and producer service sectors. The centres of some former manufacturing cities have been physically transformed through massive investment in convention centres, offices, hotels and retail and leisure developments (Sassen 1994: 43). Whether this has provided a sustainable basis for the economic regeneration of cities has looked increasingly precarious in the face of heightened inter-urban competition and problems such as the collapse of the commercial property market in the late 1980s. Other centres, having failed to harness any mechanisms of economic dynamism, have declined along with the status of their cities. Historic cities have relied on their traditionally strong tourist and visitor economies to stay vibrant (Page 1995). However, the story of the inner-city has been one of almost total and general decline. 'Inner-city' conditions and problems have also become reproduced on a number of peripheral municipal housing estates in cities like Liverpool and Glasgow. The future growth areas appear to be suburban or ex-urban, new metropolitan spaces beyond the boundaries of existing cities, discernible, for example, in the counties surrounding Los Angeles in California. The future of the economic geography of the city appears to be one of increasing decentralisation mediated through transport and telecommunication advance and change surrounding an inner-city becoming progressively disengaged from the formal economy. This is what is meant by the 'urban doughnut'.

Conclusions

This chapter has highlighted, at the broadest scale, the relationship between macro-economic change and urbanisation. It has also outlined some of the outcomes of the new micro-economic geographies of the city. Of major importance to urban geography are the ways that cities have responded and tried to combat the negative impacts of economic change. These responses have been diverse and their implications wide-ranging. Attention turns in the chapters that follow to these changes and their impacts on the landscapes, economies, images and social geographies of the city.

Further reading

Good discussions of the rise, significance and the characteristics of global cities can be found in:

Allen, J. and Hamnett, C. (eds) (1995) *A Shrinking World? Global Unevenness and Inequality* Oxford: Oxford University Press (Chapter 3)

Johnston, R.J., Taylor, P.J. and Watts, M.J. (eds) (1995) *Geographies of Global Change: Remapping the World in the Late Twentieth Century* Oxford: Blackwell.

Knox, P.L. and Taylor, P.J. (eds) (1985) *World Cities in a World System* Cambridge: Cambridge University Press.

Sassen, S. (1991) *The Global City: New York, London, Tokyo* Princeton, NJ: Princeton University Press.

Sassen, S. (1994) *Cities in a World Economy* Thousand Oaks, CA: Pine Forge Press.

An interesting account of the impact of telecommunications on the urban economy can be found in:

Graham, S. and Marvin, S. (1996) *Telecommunications and the City: Electronic Spaces, Urban Places* London: Routledge (Chapter 4).

⬤5 Urban policy and the changing city

- ⬤ Central–local government relations
- ⬤ The central government and urban regeneration – 1980s and 1990s
- ⬤ American urban policy
- ⬤ Entrepreneurial local governance – characteristics and theories

This and the two chapters that follow deal with responses to the urban economic problems outlined in earlier chapters. This chapter examines the political responses, the ways in which central and local government have reacted and the ways they have been shaped by these responses. It looks at the tensions between central and local government that arose over the restructuring of the latter in the 1980s, particularly in the UK, and the ways that local authorities have become more entrepreneurial, forming, for example, an increasing number of partnerships with the private sector. The chapter concludes with a brief discussion of theories of urban governance.

The urban problem in the 1980s

The two biggest urban, economic problems of the early 1980s in both the UK and the USA were a massive increase in unemployment and widening social polarisation. Unemployment was felt most heavily in the inner areas of older industrial cities among groups such as the young, the poorly qualified, semiskilled male workers and ethnic minorities. Opportunities were shaped by the emerging employment structure characterised by a high degree of polarisation between a small number of highly paid jobs requiring good qualifications and a larger number of very poorly paid jobs, servicing the former sector. There was no mechanism linking the two sectors and many were disengaged from the formal employment sector.

The academic and popular media have talked of the emergence of an urban underclass in cities of the UK and especially the USA in the 1990s. This was something of an oversimplification, which disguised the divisions among the less well off in British cities. Two of the most widely debated underclasses were first the working-class communities in former industrial areas, left stranded with problems of long-term male unemployment and problems by deindustrialisation, and second, ethnic minorities in inner-city areas. One of the most common metaphors that emerged to describe the social polarities in British cities in the 1980s was the 'divided city' or 'dual city'.

Since 1980 British urban landscapes, especially those of city centres, have been heavily influenced by developments in the USA. Similarly, there was a definite exchange of ideas on urban policy issues between the UK and the USA in the 1980s (Hambleton 1995a). The broad philosophy of urban policy in the two countries, as well as a number of specifics, bore a close resemblance. The private sector was targeted in both countries as a main agent in the alleviation of urban problems. This interchange of ideas on urban policy between the UK and USA appears to be an ongoing, enduring aspect of urban geography. This section will look at British urban policy in the 1980s and, drawing on policies from the UK and the USA, will briefly consider how it has evolved in the 1990s.

Broadly, urban policy since 1980 can be divided into two periods roughly corresponding to the 1980s and 1990s. Urban policy during the 1980s, in both the UK and the USA, was primarily characterised by a targeting of the private sector in partnership to produce property-led models of urban regeneration. Towards the end of the decade this drew widespread criticism on two fronts. First, there was some doubt about the ability of this approach to alleviate underlying social and economic problems that had dogged cities since they began to deindustrialise; in some cases it was felt they actually exacerbated social exclusion. Second, the array of urban programmes appeared both fragmentary and confusing. With these limitations revealed, urban policy in the 1990s aimed to become more socially inclusive, involving community interests as well.

Urban problems have always tended to be high on the media and political agenda. This was certainly the case in the UK during the 1980s, with several civil disturbances in inner-city areas such as Handsworth (Birmingham), Brixton (London) and Toxteth (Liverpool) in 1981 and 1985. The notion of the 'sick city' was further confirmed in the mind of the national media by the emergence of problems such as drug abuse, car theft and ram-raiding on peripheral housing estates in cities like Newcastle upon Tyne, Glasgow and Coventry in the early 1990s. Being seen to address these problems was a political priority for the Conservative Party for much of its term of office between 1979 and 1997. The Conservative government launched a wide-ranging urban programme in the 1980s. One of the main aspects of this was a reform of local government.

Restructuring local government in the 1980s

The period from 1945 to 1970 was characterised by steady growth in the UK economy and by the development of an extensive state welfare system. This was supported, on the whole, by both central and local government. However, this consensus began to break down in the 1970s with the rise of a political ideology termed the 'New Right'. The New Right challenged the assumption that the state was the most efficient and equitable supplier of welfare services, and argued that these could be better supplied by the market.

The New Right ideology came to dominate central government thinking after the Conservative Party election victory in 1979. Under this administration a sweeping reorganisation of local government occurred. Local government was perceived as being overly bureaucratic and an unnecessary impediment to the operation of the market. The restructuring of local government that ensued was under the banner of 'freeing the market'. It had two primary aims: reducing the power and independence of local government and bypassing local government altogether by imposing central government agencies on key policy areas directly in urban areas. To these ends the Conservative government passed some fifty Acts between 1979 and 1989 (Riddell 1989: 177; Goodwin 1992: 78). This has led to the power and independent status of local government being reduced in four key areas: local finance, the provision of services, the loss of autonomy over those services that have remained in the public sector and urban regeneration (Goodwin 1992).

This restructuring has had a profound effect on the independence of local government, the scope of its operation and the extent of local democratic accountability. Both the independence of local government, which has traditionally been greater in the UK than much of the rest of Europe, and the scope over which it has control have been reduced. Control and provision of many activities previously delivered by local government have been centralised or have been subject to the control of non-elected local agencies or moved into the private sector (Goodwin 1992; Ambrose 1994).

Many agencies, such as Urban Development Corporations, have no formal level of local democratic accountability and are answerable only to central government (Imrie and Thomas 1993a). Similarly many non-elected agencies, such as regional health boards, have no obligation to hold public meetings. The terrain of local government has progressively left the public domain since 1980. This has included key areas of local government such as health, education, and some major planning issues. Public and local representation on the boards is typically low, while representation from the business world has increased (Imrie *et al.* 1995).

> [The] central state's restructuring of local government is reflected in the emergence of centrally appointed and directed agencies with responsibilities for many aspects of the governance of the British cities. This is typified by a plethora of organisations like the Urban Development Corporations, Enterprise Agencies, Training and Enterprise Councils, in England and Wales, and Local Enterprise Councils in Scotland. They signify a new future for cities, including the removal of key powers from elected local government, the closure of their board meetings to members of the public and the press, the non-disclosure of the resultant minutes and records, and the pursuit of a politics of growth which seeks to enhance the powers of private sector interests in local economic development.
>
> (Imrie *et al.* 1995: 32)

One key area that has been a constant focus of central government attention since 1980, and which has both reflected and affected the restructuring of local government, has been urban regeneration.

Central government and urban regeneration

The Conservative government passed a range of measures during the 1980s and 1990s with the intention of targeting run-down inner-city areas – particularly those that suffered from the effects of deindustrialisation – to facilitate their regeneration. Despite the diversity of these programmes they reflected the concerns of the New Right to bypass what it saw as the bureaucratic impediment of local government. Its aim was to encourage the regeneration of these depressed urban areas by regenerating the economic infrastructure and implanting private sector mechanisms. Some of the most prominent of these programmes have been a series of grants for urban regeneration, the Enterprise Zone programme and the Urban Development Corporation programme.

Grants for urban regeneration

These constitute grant aid targeted at the regeneration of inner-city areas. The three regeneration grants available were the urban programme, the city grant and the derelict land grant (Table 5.1). The most significant of these was the urban programme, aimed at areas with special needs according to measures of social deprivation. The aims of this grant included enhancing the social and economic infrastructure of areas thus improving their prospects as locations for investment (B.D. Jacobs 1992: 146). Both the city grant and the derelict land grant were aimed at facilitating the regeneration of physically degraded land; the former was paid to the private sector, the latter to the voluntary sector.

Enterprise zones

Enterprise zones were relatively small areas of land in which special incentives were provided to encourage firms to locate there. These incentives included significant exemptions from local rates and certain central taxes, relaxed planning restrictions and reduced government interference. These incentives were in place for ten years following designation.

The enterprise zone policy was introduced in 1981 and early designations closely reflected the geography of deindustrialisation, being concentrated in large urban areas with declining manufacturing economies. These included the Isle of Dogs in inner London, Gateshead in Newcastle upon Tyne and Clydebank in Scotland. The second set of designations focused upon localities which were affected by the collapse of the British Steel industry. They included towns like Rotherham and Scunthorpe (Hudson and Williams 1986: 115).

One of the few dynamic sectors of the UK economy during the early 1980s was the small firms sector in areas such as computing and hi-tech manufacturing. It was these types of firms that enterprise zones were designed to attract, with the intention of establishing regional growth

Table 5.1 Expenditure on inner cities (UK £ million)

	1986–87 Out-turn	1987–88 Out-turn	1988–89 Out-turn	1989–90 Out-turn	1990–91 Estimated out-turn	1991–92 Plans	1992–93 Plans	1993–94 Plans
Urban programme	236.7	245.6	229.0	226.0	232.6	243.2	243.6	245.9
City grants	23.9	26.8	27.8	39.1	49.0	66.0	76.0	78.7
Derelict land	78.1	76.5	67.8	54.0	64.3	75.5	83.1	86.8
UDCs	89.3	133.5	234.4	436.0	550.2	473.1	305.4	302.8

Source: Department of the Environment Annual Report 1991: The Government's Expenditure Plans 1991–2 – 1993–4 (1991) cited in B.D. Jacobs (1992)

multipliers. These firms offered the potential of economic growth coupled with the diversification of the local economy. The success of the enterprise zone policy has been the subject of some debate. They certainly created new jobs, an estimated 8 000 between 1981 and 1984, many of which were in new companies. However, compared to national unemployment levels of around 3 million, this was a drop in the ocean. Despite attracting some new firms, enterprise zones failed to diversify their local economies significantly. The types of firms attracted to enterprise zones tended to mirror those already present in the surrounding area. Consequently enterprise zones could be regarded, at best, as buttresses of their local economies rather than the catalysts of their regeneration (Lawless 1986). As well as this, some enterprise zones were outright failures, for example, Belfast's enterprise zone had failed to attract 100 jobs by 1984 (Hudson and Williams 1986: 115).

A policy, such as the enterprise zone policy, which offers incentives within a spatially defined area has an inherent negative impact on the areas around it. These surrounding areas automatically become less attractive locations for investment. Enterprise zone locations offered savings on rates of up to £10 000 per annum in some cases. Adjacent areas without these savings immediately become devalued, being able to offer no compensating advantages. This is known as the 'boundary' or 'shadow' effect (Norcliffe and Hoare 1982; Hudson and Williams 1986: 116; Lawless 1986: 47).

Urban Development Corporations

The Urban Development Corporation (UDC) programme can be regarded as the flagship of the Conservative government's urban regeneration programme in the late 1980s and 1990s. In the amounts invested in them, the political and media attention they received and the extent to which they embodied the New Right ideologies, they surpassed all other urban programmes of the period. Urban Development Corporations are government agencies, implanted directly upon designated urban areas and responsible for the regeneration of these areas. They are run by appointed boards consisting largely of representatives from the local business community; there is typically little representation from the local public sector (Imrie and Thomas 1993a). The Urban Development Corporations aim to encourage the private sector back into run-down inner-city areas, thus rebuilding their economic infrastructures. This is achieved through a market-oriented and property-led approach. The Urban Development Corporations encourage regeneration in a number of ways acting as partners to private sector companies (see Table 5.2 and Figure 5.1).

Table 5.2 *Urban Development Corporations in the UK*

	Area at designation (ha)	Population at designation	Employment at designation	UDC grant aid 1989–90 to 1991–2 (£m)
Birmingham Heartlands	1 000	12 500	Not known	NA
Black Country	2 598	35 405	53 000	96
Bristol	420	1 000	19 500	22
Cardiff Bay	1 093	500	15 000	94
Central Manchester	187	500	15 300	41
Leeds	540	800	NA	30
London Docklands	2 150	40 400	27 213	813
Merseyside	350	450	1 500	72
Sheffield	900	300	18 000	42
Teesside	4 858	400	NA	110
Trafford Park	1 267	40	24 468	76
Tyne & Wear	2 375	4 500	40 115	110
All	17 738	96 795	214 596	1 506

Source: Imrie and Thomas (1993a: 13–15)

The responsibilities of Urban Development Corporations are wide-ranging. They cover the reclamation of land, the provision of transport infrastructure, environmental enhancement and the provision of financial incentives to the private sector (Imrie and Thomas 1993a). Urban Development Corporations can be considered facilitators of regeneration; their designation lasts for between five and fifteen years. Thereafter, it is intended that they be disbanded to let the economic mechanisms they have helped to implant continue the long-term regeneration of their areas.

Urban Development Corporations have proved an extremely controversial measure and have drawn heavy criticism on a number of economic, political, social and cultural fronts. Despite massive government expenditure they appear to have generated very few jobs. Most of the economic activity taking place within them has been based

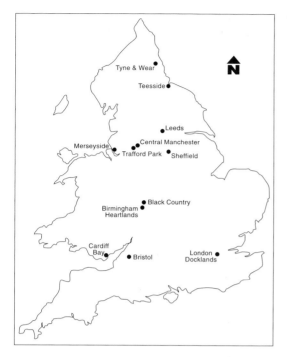

**Figure 5.1   Map of Urban Development
Corporations**
Source: Imrie and Thomas (1993a: 12)

around property speculation (Byrne 1992). Indeed, in creating an infrastructure based around office and leisure development Urban Development Corporations may actually have largely precluded or retarded future reindustrialisation of their areas (Byrne 1992: 264; Hamilton-Fazy 1993: 19). While they have generated some inward investment, this has been small compared to the public money invested in them. There has also been criticism of the ways in which this money has been invested, as evidence came to light in the early 1990s of Urban Development Corporations losing massive amounts of money on land deals (Brownhill 1990; B. D. Jacobs 1992; see also Case Study E).

Politically the Urban Development Corporations have been central to the removal of local democratic accountability in local governance and its replacement with non-elected boards and centrally accountable agencies. In taking control of key areas of urban governance such as planning they have been a regular source of antagonism to local councils (B.D. Jacobs 1992; Imrie and Thomas 1993a; see also Case Study F).

Social tensions and conflicts have also emerged within areas controlled by Urban Development Corporations. It has been common for working-class communities to express their opposition to what they see as their exclusion from the development and local decision making processes. They have regularly voiced fears about being physically displaced by extensive developments or socially displaced by more wealthy incoming populations (Brownhill 1990; G. Rose 1992). The London Docklands Development Corporation has long been subject to such opposition (Plate 5.1). These criticisms are explored in Chapter 8.

Case Study E

Urban Development Corporations and land deal losses

Stephen Byers, Labour MP for Wallsend, reported that Urban Development Corporations have lost a total of over £82 million of public money in unsuccessful land deals. A comparison of the prices paid and current values of land revealed the losses. The land purchased by the Urban Development Corporations was bought at the height of a property boom. However, since then land prices have fallen sharply as economic recession halted the expansion and relocation of companies (Table 5.3). Stephen Byers has argued that this provides evidence that land speculation and property development are not successful routes to economic regeneration.

Table 5.3 *UDCs and land deal losses*

	Amount spent on land transactions (£m)	Amount received from land transactions (£m)	Value of land at 31 March 1992 (£m)	Loss (£m)
Black Country	68.0	3.7	38.8	26.5
Bristol	24.4	nil	11.5	12.9
Central Manchester	12.3	nil	4.0	8.3
Leeds	17.5	1.8	12.3	3.4
Trafford Park	41.6	9.3	22.3	10.0
Tyne & Wear	42.4	5.8	30.3	6.3

Source: Wynn Davies (1992: 2)

Central government urban policy in the 1980s: an assessment

In 1989 the Audit Commission published a report, *Urban Regeneration and Economic Development*, which provided a review of British urban policy since 1979. The report was critical of a number of aspects of these policies. The most serious included:

- the plethora of government departments responsible for urban policy
- the complexity of the patchwork of different urban policies
- the conflicts between central and local government over urban policy
- the perception of local authorities that they have become excluded from the urban policy arena

Plate 5.1 *Poster from London Docklands Community Poster Project*
Source: Peter Dunn and Loraine Leeson, 'Big money is moving in', in London Docklands Community Poster Project,
The Changing Picture of Docklands 1981–5 series, 18 ft × 12 ft photomural, B/W montage, hand coloured with
acrylic glaze, by permission of The Art of Change

- the failure in operation of many partnerships schemes linking the
  public and private sectors.

In response the Commission concluded that urban policy in the 1990s
should attempt to overcome these problems. They argued that it should
be more straightforward and less complex and that local authorities
should take a more active role in policy formation and instigation (B.D.
Jacobs 1992: 144–5).

Urban policy in the 1990s

The emphasis of urban policy shifted somewhat in the 1990s. Policies
moved away from two-way partnerships between the public and the
private sectors, based on property-led regeneration and the 'trickle-down'
of investment. Rather they sought to foster a three-way partnership model
which included community organisations and the voluntary sector as well
as the local authority and the private sector. Indeed the rediscovery of the

Case Study F

Urban regeneration in Bristol

Bristol's Urban Development Corporation was designated on 7 December 1987 by the Secretary of State for the Department of the Environment. Despite being in one of the fastest growing regions in the UK, which enjoyed growth in sectors like insurance, banking and finance, medical services and hotels and catering, a number of economic problems had begun to emerge in Bristol in the late 1980s. The main problems revolved around job losses in sectors including paper and packaging, food, drink and tobacco and the defence industries. These had previously been staples of the regional economy.

In 1986 the Department of the Environment had recognised four problems that the Urban Development Corporation was later designated to tackle:

- growing polarisation in the labour market
- spatial polarisation in the city and the development of areas of high deprivation
- ongoing problems of social order
- tensions between a Labour-controlled local authority and the Conservative central government.

Specific barriers to regeneration that the Urban Development Corporation was charged with addressing included problems with derelict land and premises, an ageing and inadequate road system, land contamination and fragmented patterns of land ownership.

The designation of an Urban Development Corporation in Bristol drew immediate and sustained criticism from the local authority, which expressed concerns about five issues: outside intrusion into the key areas of planning and development, the use of public funding to subsidise private development, the lack of the Urban Development Corporation's local accountability, potential clashes with existing local government policies and the definition of regeneration within the designated area. The local authority felt that the Urban Development Corporation was more concerned with the regeneration of the local property market than with local social needs. The result of these criticisms was that the City Council challenged the urban problems that the Department of the Environment had highlighted and argued that there was no justification for the designation of an Urban Development Corporation. This challenge forced a House of Lords inquiry. While finding that the City Council had cause for complaint it eventually endorsed the designation. The inquiry did, however, make a number of recommendations; these were that four areas within the initial designation should be excluded and that the Urban Development Corporation should contain objectives linked to social as well as physical and economic regeneration and that it should not disrupt or displace existing employment. The inquiry explicitly stated that it did not wish to see the mistakes of London's Docklands repeated in Bristol.

Despite the recommendations of the House of Lords inquiry the objectives of the Bristol Development Corporation stressed the importance of market mechanisms and generating new economic activity in their plans for regeneration. The strategy was based upon the transformation of Bristol's image through prestigious architectural projects and marketing campaigns. The proposals included the displacement of around 300 jobs from the area while the Urban Development Corporation had a poorly defined relocation policy. The proposals favoured physical rather than social regeneration and there were no guaranteed mechanisms outlined to direct employment benefits towards local people.

These proposals led to fears from the local authority, local businesses and local people over disruption, displacement and contravention of local plans. There emerged a polarisation of opinion between local interests who wished to maintain and build upon the local economic structure and the Urban Development Corporation who believed that restructuring the local economy was the key to regeneration.

The fears of the local interests were not allayed by the Urban Development Corporation's processes of consultation. Consultation with the local authorities and communities was ad hoc and proved to be more an exercise in persuasion than a mechanism to facilitate genuine participation in the development process. The consultation resulted in no substantial response from the Urban Development Corporation over the fears raised by the local people or the local authority. The Urban Development Corporation decided that no changes to their proposals were needed despite a number of specific concerns being raised. The Urban Development Corporation seemed more willing to consult with local businesses and land and property interests than with the local people and authority. These sessions were aimed, again, more at winning over support than involving local interests in development.

Ultimately the acid test for the Urban Development Corporation would be its successes or failures to regenerate the area. By 1992 little progress had been made to this end. Many sites within the urban development area were still derelict and the development process had become bogged down in a number of costly and lengthy public inquiries. The recession of the late 1980s, which particularly affected the property market, had demonstrated the limitations of relying on market mechanisms and private capital for regenerating local economies. The lack of alternative strategies precluded any effective response to the prevailing economic conditions.

Source: Oatley (1993)

community has been trumpeted as the success story of urban policy in the 1990s (Davoudi 1995; Hambleton 1995a). The two most significant urban policies of the early 1990s have been City Challenge, announced in May 1991, and the Single Regeneration Budget (SRB) launched in November 1993. Such policies have aimed to incorporate local residents, especially in 'disadvantaged' neighbourhoods, into the decision-making processes that affect their local area, or to empower them.

A major difference between these programmes and those of the 1980s is that they were aimed at creating more sustainable forms of regeneration. First, they aimed to be catalysts that created new structures and practices which might be replicated elsewhere. Second, the institutional capacity, formed by the programmes, the structures, partnerships and arrangements that they generated, was as important as the actual material outcomes of the policies. It was this capacity, once created, that was deemed likely to sustain regeneration into the future. A further aim, especially of City Challenge, was to change the way local government operated, making it less bureaucratic and more responsive to both the market and to the needs of its local community (Davoudi 1995: 333–4). Despite these qualitative aims both City Challenge and SRB contained a number of quantitative economic and environmental aims.

Funds for both of these policies were, and continue to be, allocated through competition. In the first round of City Challenge bidding, of fifteen local authorities invited to bid, thirteen were successful. They each received £7.5 million per annum for five years. A second round of bidding produced twenty successful authorities out of fifty-seven invited authorities (Shiner 1995: 39).

The SRB was a response to the criticism that the framework for urban regeneration programmes in the 1980s was fragmented and confusing. SRB pulled together twenty budgets which were previously administered by several departments. A total of £1.2 billion has been set aside for the SRB which breaks down to between £100 million and £140 million per annual round of bidding (Shiner 1995: 39). Despite representing a major shift away from the much criticised models of urban regeneration of the 1980s, both programmes have also been subject to a number of major criticisms.

City Challenge was vaunted as a means by which new voices would be introduced to the arena of local governance and governance would be transformed to make it more responsive to these interests. While it is true that this is a policy that has broadened access to local political power it has tended to perpetuate certain existing inequalities in this access. Because of the haste with which many initial City Challenge bids were produced, local authorities tended to rely heavily on existing local community and business networks and agencies (Davoudi 1995: 342–3; Hambleton 1995a). These were far from comprehensive in the areas and groups they included. This local authority reliance tended to exclude areas with poorly developed community organisation and small firms, and some larger corporations, not part of business networks or agencies. Similarly, City Challenge has so far failed to have much effect on the operation of local government. Rather than transforming local government, the operation of City Challenge has been shaped by the existing operations and procedures of it. In addition to this City Challenge, progress has been adversely affected by antagonisms between the partners in the three-way partnership and the effects of economic recession, especially on inner-city job markets (Davoudi 1995: 341).

The SRB has been subject to a different set of criticisms. The most serious of these surround the criteria used for the allocation of funding and the actual amounts of funding available. The SRB has replaced a number of existing budgets including the urban programme. This has effectively reduced deprivation as a criterion for fund allocation. This has produced some startling allocations which have failed to include a number of severely deprived areas which received funding under previous programmes. For example, while areas such as Leicester, Nottingham and Walsall have failed to receive SRB funds, Bedford, Eastbourne, Hertfordshire and Northampton have benefited from around £17 million of funding. The SRB has also been accused of being a means by which central government has disguised the fact that the rationalisation of urban regeneration budgets has actually reduced the amount of central government funding available and is likely to continue to do so (Shiner 1995: 39).

The aims of urban regeneration programmes in the 1990s have been to produce a more self-sustainable form of regeneration, to broaden access to power and to transform the operations of local government. While it is too early to provide a fully rounded evaluation of these policies, initial signs do not appear overwhelmingly encouraging. The policies appear hamstrung by the legacy of the policies of the 1980s and existing local government structures. These have exerted a disproportionate influence over the direction of policy rather than the opposite, intended effect (Davoudi 1995).

Future urban policy: lessons from the USA?

US urban policy, as evolved under the Clinton administration, has been generally better received than the versions implemented in the UK, particularly with regard for its emphasis on community regeneration as a central aspect of policy. US urban policy is essentially two-pronged: the Empowerment Zones programme has created nine designations (six urban and three rural) and the Enterprise Communities programme ninety-five (sixty-five urban and thirty rural) (Hambleton 1995a: 363; 1995b: 29). These designations were announced in December 1994, the result of bids by 500 applicants. The Empowerment Zone distribution reflected the problems concentrated in the formerly industrial cities of the north-east: New York, Philadelphia/Camden, Baltimore, Detroit (see Case Study G), Chicago and Atlanta. These problems are not exclusive to these cities but are found in other areas throughout North America. The failure, for example, of Los Angeles to win Empowerment Zone status was a major cause of resentment within the city (Hambleton 1995a: 367). In total the urban Empowerment Zones received US $600 million, the Enterprise Communities US $285 million and a further six areas received a total of US $315 million under additional designations as Supplemental Zones and Enhanced Economic Communities (Hambleton 1995a: 366). Along with federal investment, designated areas received tax incentives to help create employment and were expected to have secured a commitment from the private sector for investment.

The main distinguishing features of Empowerment Zones are that they directly target disadvantaged populations and that the community is central to the planning and implementation of Empowerment Zones initiatives (McCarthy 1995: 347). One of the four guiding principles of the Empowerment Zone policy is that the vision contained within the bid should be based on widespread local support (Hambleton 1995a: 364). Some British politicians have shown enthusiasm for initiatives within the Empowerment Zone programme (Hambleton 1995b: 29).

Local authorities and economic development

The focus of local government activity has shifted since the early 1980s away from issues of social welfare towards those of economic development. It has been argued that this represents a shift away from managerial modes of governance towards entrepreneurial modes of

Case Study G

Detroit Empowerment Zone Plan

Detroit's inner areas, like those of many other large cities in north-east USA, are characterised by high levels of social and economic disadvantage. For example, within the 18.35 square miles of the designated Detroit Empowerment Zone, 47 per cent of the population are in poverty (higher for children), the median family income is half that of the Detroit metropolitan area and over half of the adults over 25 years of age do not have a high school diploma. The area also suffers problems linked to low rates of car ownership, poor health, high rates of alcoholism and homelessness. However, it was these very problems that made Detroit eligible to apply for Empowerment Zone status from the US government. The Empowerment Zone policy is based on the principle of public bodies, the private sector and networks of community organisations working together to formulate plans that gain local support through participation and local accountability. The formulation and implementation of Detroit's plan is a good example of the type of capacity that can be created by this approach.

In regeneration terms the Empowerment Zone policy has two main aims. The first is to secure private funding for the area, but with the intention of achieving the second aim: reconnecting areas and communities with formal opportunities for improvement. The disarticulation of neighbourhoods and people from formal economic mechanism has been one of the most worrying of recent urban problems.

Community participation was central to both the formulation and implementation of Detroit's bid. Elected community representatives comprised nine of the eighteen panel members who selected the area for the application, formulated the strategic plan and chose specific projects for funding. The implementation and running of Zone initiatives is done by a fifty strong Empowerment Zone Development Corporation, thirty of whom are community representatives.

Empowerment Zone status has brought with it a number of public and private financial incentives. These include a $100 million federal block grant to be used for improvements to the physical and social infrastructure, tax incentives to employers who recruit labour from inside the Empowerment Zone and a pledge of $1.9 billion of private funding. Of the latter, $1 billion will form a loan fund administered by thirteen local financial institutions. This will make loans available to small businesses and potential home buyers who might have trouble otherwise securing credit.

The Detroit plan includes eighty projects aimed at improving training provision in the zone, regeneration and social housing. One example is the 'One Stop Capital Shop', a centre aimed at assisting small businesses with information and planning. Many of these projects are run by agencies on a 'not-for-profit' basis.

Detroit's Empowerment Zone plan appears to offer a model of an inclusive, bottom-up, more people-centred model of regeneration. It has been characterised by strong political leadership in conjunction with local partnership and consensus, a stark contrast with the political and communal divisions that have surrounded many British urban regeneration projects. While US urban policy is far from being immune to criticism, if the Detroit plan is implemented as successfully as it appears to have been planned, it is likely to offer a model of urban regeneration that involves more than simply 'paving the way for private sector investment' (Hambleton 1995a: 373).

Source: McCarthy (1995)

governance. This is something of a generalisation which ignores, for example, the enormous diversity of local government regimes within and between countries (T. Hall and Hubbard 1996). However, it is useful in that it marks out important trends in urban politics.

The entrepreneurial phase of local government has coincided with the limitation and the decrease in central government funding to local government and the restriction of the amounts that local governments are able to generate through local taxation. This has been on top of the severe deindustrialisation of many urban economies. The result has been a severe drop in the funding available to local government (Goodwin 1992).

This situation has forced local authorities to adopt a much more entrepreneurial role, generating revenue through functions which were traditionally the preserve of the private sector. These functions have included speculative property development, place promotion and subsidy of private development. Many agree that this increased involvement of the public sector in initiating economic development represents a major shift in the nature of local government. Local authorities have formed an increasing number of alliances with outside interests to further their aims of local economic development. These other interests, predominantly from the private sector, have also found advantage in these alliances. This might include preferential treatment from local authorities, for example. The relationships between local authorities and other interests within their urban area have generated a great deal of literature from urban geography, political science and sociology. This has led to a number of theories that have sought to explain the nature of urban politics in terms of who governs cities, who has access to power, and who is excluded from power? The section that follows will consider two of the most influential of these theories: growth coalition theories and regime theories.

Growth coalition theories

A number of North American cities have long been characterised by a variety of alliances between groups such as land owners, rentiers, state banks, utility companies and local authorities. These alliances have typically been formed with the intentions of intensifying urban development and attracting outside investment. These outcomes are to the advantage of both private interests and the local authority (Logan and Molotch 1987; Cox and Mair 1988). Many observers have argued that private interests achieve their aims by becoming influential within local politics. This is often achieved by convincing the local population that such developments are to the benefit of the city as a whole rather than just a small economic elite (Thomas 1994; T. Hall and Hubbard 1996). The composition and workings of growth coalitions are easily explained with reference to the role of the local media within them.

The local media usually play a key role within such coalitions. Their coverage of major urban development is typically uncritical and boosterist. Their analysis is rarely in depth and they tend to give precedence to local authority sources rather than alternative or oppositional sources. The local media gain from a potential increase in both circulation, if the city grows, and advertising revenue. In supporting the local authority and developers they maintain regular and plentiful supplies of easy news (Thomas 1994).

Initially writing in this vein emanated from the USA and concentrated on some famous examples of the involvement of local government in economic development. This involvement was often led by charismatic individuals such as mayors, councillors and other politicians.

Application of growth coalition theories to the UK

Appealing as this growth coalition theory may appear, there are a number of problems when applying it to UK cities (Boyle and Hughes 1995). UK and US local government varies in a number of key ways. The role of local authorities in urban regeneration can be dated back to the nineteenth century. This was stepped up and focused more specifically on economic development in the 1960s. This entrepreneurial activity appeared to intensify during the 1980s. A number of local authorities in the UK, especially in older industrial cities, appeared to be acting in a more entrepreneurial way. For example, Sheffield City Council subsidised the development and hosting of the World Student Games in 1992; this left the city in severe financial crisis afterwards. Strathclyde Regional Authority and Glasgow District Authority invested heavily in physical development and marketing to support Glasgow's role as European City of Culture in 1990. Birmingham City Council invested more than £100 million of public money in the International Convention Centre which opened in 1991, and continued to underwrite its losses for a number of years afterwards. However, the extent to which this represents a fundamental shift in the nature of local governance in the UK is a matter of some debate. There are a number of examples of local authorities pursuing no-growth, limited-growth or anti-growth strategies. These include towns like Canterbury and Cheltenham (T. Hall and Hubbard 1996: 158). This is usually the case where development poses a threat to an historic townscape. Rather than simply talking in terms of a general shift from managerial to entrepreneurial modes of local government it is more helpful to recognise the diversity that exists between different authorities and to recognise the importance of geographical variations in local economies and land and property markets, the nature of the local urban fabric, the composition of local agencies, the political biases of local authorities and their relationships with their local business communities. Factors such as these appear to have had a significant influence upon the determination of the course of local authority action between the apparently competing imperatives of social welfare provision and economic development.

Criticisms of growth coalition theories

While they highlight some of the important aspects of the relationship between urban politics and economic development, growth coalition theories do have their limitations. Two major related weaknesses are their inability to account successfully for diversity between coalitions and the naive way that they conceptualise the issue of power in urban politics (Stoker 1995). Growth coalition theorists argue that the prime motivations behind coalition formation are economic or political (to bring more money into an area or to win more votes). Consequently the picture it tends to paint of the urban political arena is very polarised, dominated by groups who are either pro- or anti-economic development. The classic picture is of a business community that is strongly pro-growth and a public that is very anti-growth. As a result growth coalition theories are unable to account successfully for the public as active members of growth coalitions (Stoker 1995). However, a number of studies in the USA have shown that the support of the public is a crucial factor in coalitions achieving their objectives. The motivations that have lain behind coalition formation and the groups that have taken an active part have been far more diverse than growth coalition theories have been able to take account of.

Much of this weakness stems from a very narrow and unproblematic conception of power within growth coalition theories. Growth coalition theories regard power in urban politics as the ability to control people and resources. While this might ultimately be the case, it fails to acknowledge the importance of the processes whereby this power is created. An alternative set of theories called regime theories focus specifically on the complex processes in the creation and the production of power (Stone 1989; Stoker 1995). In doing so they offer a much more sophisticated account of the operation of urban politics.

Regime theories

Regime theories start from the observation that urban politics is characterised by a fragmented distribution of power (see Case Study H). The actual power that is held by individual groups, be they governmental or non-governmental, is in reality very limited. They each control only segments of society and economy. Regime theory focuses on the ways that groups in urban politics overcome their own inherently limited power by coming together and forming regimes to achieve specific objectives. The crucial contribution of regime theories is that they demonstrate that the power to govern is not *given*, it is not the inherent possession of any group; rather, it has to be *created* or *produced*. It is created, regime theorists argue, by different groups coming together and blending their control and resources. In doing so they are able to create the capacity to govern (Stoker and Mossberger 1994; Stoker 1995: 269–70).

Typically regimes are formed through some combination of a government organisation that has the ability to mobilise and co-ordinate resources and private sector interests who might own resources. Often otherwise opposed groups are able to come together by each partner offering selective incentives to the other. Sections of the public can enter regimes by offering their support for developments either through the ballot box or expressed through community leaders (Stoker 1995). Regime theory is, therefore, able to transcend a major weakness of growth coalition theory which is able to conceive of the relationship between certain groups only in a very polarised, oppositional way (Painter 1995; Stoker 1995). Regime theory is able to show how, while this opposition may be constraining, it does not necessarily determine the course of urban politics. Groups may be able to transcend their divisions in specific instances if, in doing so, they can achieve a capacity to govern. Consequently, regime theory paints a picture of urban politics as less rigid and structurally determined than growth coalition theory and regime formation as more fluid and complex (Stoker 1995).

Regime theory is a relatively new addition to the study of urban politics. Although far from being the finished article it has proved an increasingly influential contribution to the geographical studies of urban politics and economic development (Judge *et al.* 1995). Although developed in the context of North American cities it appears to be more readily applicable to the British and European political arenas than cruder, growth coalition

Case Study H

Different types of regimes

Urban politics is characterised by a diversity of types of regimes, which are composed of different groups with very different aims. Four main types of regimes have been recognised from studies of American cities. Maintenance Regimes seek the preservation of the status quo rather than the promotion of new development. Development Regimes seek to promote development or to arrest or prevent decline; this task requires greater resources and co-ordination than that facing maintenance regimes. Middle-Class Progressive Regimes may seek a variety of outcomes from development including social

gains; they may also seek to control development to prevent or limit externalities such as environmental damage. Lower-Class Opportunity Expansion Regimes seek to enhance or expand the opportunities open to disadvantaged urban groups, which typically involves major resources and substantial co-ordination. The scale of these tasks and the relatively disempowered positions of the groups in this type of regime means that it is often absent from American cities.

Source: Stone (1993)

theories which rely far more on economic motivation to explain coalition or regime formation. Regime theories also appear far more versatile, being able to explain the actions of central government agencies (such as Urban Development Corporations) as well as local public and private agencies (Stoker 1995; T. Hall and Hubbard 1996).

Further reading

Two very useful summaries of key debates in the study of urban politics can be found in:

LeGates, R. and Stout, F. (eds) (1996) *The City Reader* London: Routledge (Part 4).

Savage, M. and Warde, A. (1993) *Urban Sociology, Capitalism and Modernity* London: Macmillan / British Sociological Association (Chapter 7).

An excellent introduction to recent advances in the field of comparative urban politics can be found in:

Judge, D., Stoker, G. and Wolman, H. (eds) (1995) *Theories of Urban Politics* London: Sage.

A number of excellent case studies on Urban Development Corporations in the UK are included in:

Imrie, R. and Thomas, H. (eds) (1993) *British Urban Policy and the Urban Development Corporations* London: Paul Chapman.

Discussions of British urban policy can be found in:

Atkinson, R. and Moon, G. (1994) *Urban Policy in Britain: The City, the State and the Market* London: Macmillan.

Blackman, T. (1995) *Urban Policy in Practice* London: Routledge.

6 ▸ Urban change and emergent urban forms

- The modern/post-modern city debate
- The production, regulation and consumption of urbanisation
- Recent urban change/new urban forms

The transformation of the city

Much has been written in recent years of the apparent transformation of the form and type of cities in Europe and North America. Much of this debate has focused on the emergence of 'post-modern', 'post-industrial' or 'post-Fordist' urban forms. Post-modern urban form is significantly different in its form, its patterns of land values, its social geographies and its landscapes to the modern city described in models such as Burgess's concentric ring model (1925) and Hoyt's sector model (1933). Such cities, which developed over the course of the twentieth century, typically displayed homogeneous zones of land-use and social group, land values which declined regularly away from the centre of the city. Urban geographers since the early 1980s have argued that this idea of the city is outdated and that we have been witnessing the emergence of the new urban forms. Despite a number of differences between individuals they generally agree that these new cities are more fragmentary in their form, more chaotic in structure and are generated by different processes of urbanisation than earlier cities. This new urban form has been nicknamed the 'galactic metropolis' (Lewis 1983; Knox 1993). This describes a city which, rather than being a single coherent entity, consists of a number of large spectacular residential and commercial developments with large environmentally and economically degraded spaces in between. They are said to resemble a pattern of stars floating in space rather than the unitary metropolitan development growing steadily outward from a single centre.

Some views of the post-modern city

> The old idea of the city focused only on the picture postcard landmarks
> and the central crust of buildings and spaces. But it is clear that the
> present-day city has long-since outstripped those limits. The new
> incarnation of the city is an endless amorphous sprawl, with which
> outcrops of skyscrapers or vast shopping malls can appear almost
> anywhere.
>
> (Sudjic 1993)

> Pushing one's face against the glass, one could see all that any human
> being could reasonably bear of St Louis: mile after mile of biscuit
> coloured housing projects, torn-up streets, blackened Victorian factories
> and the purplish, urban scar tissue of vacant lots and pits in the ground. It
> was the waste land. . . . Beside me the conventioneers were identifying
> another city altogether. They were pointing out the fine new home
> stadium of the Cardinals, Stouffer's Riverside Towers, the tall glass office
> blocks. To me the isolated sprouts of life in the surrounding blight were
> just objects of pathos: a few wan geraniums planted on a rubbish heap
> don't make a garden.
>
> (Raban 1986: 329–30)

The modern-post-modern city debate: a summary

The debate about the apparent transition from modernity to
post-modernity in society is extensive and has generated an enormous
volume of literature from architecture and urban studies to film, literary
criticism and fashion. This section aims to summarise the main
characteristics of the debate that apply to the city. This section should
be used only as a crude, short-hand guide to the debate. The actual
nature and extent to which modern forms of urbanisation have been
supplanted by post-modern forms will vary enormously between cities.
The outcome of the interrelationships between these two processes will
be unique in every case. Not all cities, for example, could be said to be
modern or industrial. Many, such as York and Durham in the UK, have
retained much of their pre-industrial structure. Likewise, in the face of
post-modern forms of urbanisation, many cities are likely to retain
much of their modern or industrial structure. In reality many cities will
demonstrate some combination of modern urban characteristics mixed
in with newer post-modern urban forms. For example, recently
redeveloped docklands areas might be surrounded by large areas of

inner-city little affected by post-modern processes. In most cases the overall structure of the city still reflects modern urbanisation processes of industrial capital and planning. However, within this largely modern structure new urban forms have begun to emerge and there is evidence that the internal space of the city has begun to be resorted or reorganised (Cooke 1990: 341).

The main characteristics of the modern-post-modern debate with regard to the city are summarised below.

Urban structure

Modern

Homogenous functional zoning; dominant commercial centre, steady decline in land value away from centre.

Post-modern

Chaotic multinodal structure; highly spectacular centres; large 'seas' of poverty; hi-tech corridors; post-suburban developments.

Architecture, landscape

Modern

Functional architecture; mass production of styles.

Post-modern

Eclectic, 'collage' of styles; spectacular; playful; ironic; use of heritage; produced for specialist markets.

Urban government

Modern

Managerial – redistribution of resources for social purposes; public provision of essential services.

Post-modern

Entrepreneurial – use of resources to lure mobile, international capital and investment; public and private sectors working in partnership; market provision of services.

Economy

Modern

Industrial; mass production; economies of scale; production-based.

Post-modern

Service-sector based; flexible production aimed at niche markets; economies of scope; globalised; telecommunications based; finance; consumption oriented; jobs in newly developed peripheral zones.

Planning

Modern

Cities planned as totalities; space shaped for social ends.

Post-modern

Spatial 'fragments' designed for aesthetic rather than social ends.

Culture and society

Modern

Class divisions; large degree of internal homogeneity within groups.

Post-modern

Highly fragmented, lifestyle divisions; high degree of social polarisation; groups distinguished by their consumption patterns.

It should be apparent that these new urban forms do not simply appear for no reason; they are the visible outcome of a whole series of complex economic, political, social and cultural processes. It is vital that the urban landscape must be read within this context. This is so for two reasons.

First, descriptions can provide only a very limited understanding of the urbanisation process. They say nothing of the underlying processes that created them. It is important to specifically examine these processes in conjunction with descriptive accounts.

Second, to say that the city reflects changes in these processes, while being true, reveals only one-half of the relationship. Changes in international economics, politics, society and culture are not simply stamped on the city without resistance. Rather, at the local level their imprint is uneven and contested. They are frequently opposed, resisted, misread, or encouraged by institutions, agents and social groups in cities. The tensions between the operation of these processes and local groups results in tension that affects the outcomes of these often global processes at the local level. Therefore, the city not only reflects the nature of the processes of urbanisation, but also is active in affecting them.

Transformation of the city: an evaluation

The debate outlined above is based primarily on a small number of cities which have become constructed as archetypes of post-modern urbanisation. These cities, primarily, although not exclusively, in North America, have included Los Angeles, New York, Washington, DC, London and Tokyo. Aspects of this debate were discussed in detail in Chapter 2. Despite the undeniable and indeed growing influence of these cities in both their national economies and the international economy, they are not necessarily representative of the experience of urbanisation in the majority of cities in the West.

Despite obvious differences these cities all possess a number of overarching similarities. Most importantly, they are all world cities or global cities, control and command points of interlinked global economies and cultures (Hamnett 1995; Knox 1995). It is, consequently, these cities that are at the hub of emergent forms of urbanisation and it is not surprising that they have been drawn upon in the construction of new models of urban form. Despite words of caution from their authors, certain assumptions underpin both the construction and use of these models. The problems of applying these models, drawn from a select group of cities, are very similar to the problems in trying to apply the models of the industrial city which were based upon the 'shock cities' of the industrial revolution, Manchester and Chicago, to general urbanisation in the West during the twentieth century.

The assumption that the processes of urbanisation shaping cities like Los Angeles and New York during the 1980s and 1990s will apply equally to other towns and cities in the West raises certain important questions of the validity of these models. It is the aim of this chapter to evaluate the applicability of the model of urbanisation outlined above. In doing so it will examine urbanisation process since the late 1970s. It is not the aim of this chapter to affirm or refute the model of urbanisation outlined above but to attempt a more subtle and hopefully meaningful critique. It will not ask: should we apply this model of urbanisation but, rather, how should it be applied and in what ways can it be regarded as a blue-print of urban change in the late twentieth century? It will ask: is it inevitable that all urban areas will be transformed in the ways that this model suggests? If this is not the case, why not? In what ways do the processes of urbanisation highlighted by the model (and discussed in other chapters of this book) feed into and change the urban hierarchy (the relations between cities) and the internal make-up of cities? What will be the spatial consequences of this?

This chapter will attempt to examine the effects of 'post-modern' urbanisation on certain generic parts of the city. It will focus on recent urban change in the city centres, the 'inner' cities, the suburbs and 'post-suburban areas' primarily in the UK. In doing so it adopts the framework for understanding urban change outlined in the introductory chapter. It will examine the ways in which urban change has been produced, regulated, and consumed since the late 1970s. This framework for exploring urban change is particularly well developed in the writings of the urban geographer Paul Knox (1991; 1992a; 1993) and is based on observation of the 'restless' landscape of and around Washington, DC. However, in situating these specific observations in a more general context, he provides a useful insight into understanding processes of recent urban change elsewhere.

The production of urbanisation

The production of urbanisation has been affected by four main changes since the early 1980s. These have involved changes in the structure of investment in the built environment, the organisation of both the development industry and the practice of architecture and the technologies employed in building provision (Knox 1992a; 1993). Together, these changes have accounted for a number of emergent forms in the landscapes of cities.

Cycles of urban development closely reflect the waxing and waning of the property markets; on occasions they represent profitable destinations for investment, at other times they do not. Urban development is likely to be more intensive when the former rather than the latter is the case.

A major wave of urban redevelopment occurred between the mid-1970s and the property recession of the early 1990s. This period transformed a number of urban landscapes. This wave of development was sparked off initially by the effects of the oil price crisis of 1973. One effect of the rise of the price of oil was a massive influx of 'petrodollars' into the economies of the West. To employ a Marxist reading of the situation, over-investment in the primary sector made this an unprofitable destination for these funds and consequently they 'switched' to the more profitable secondary (property) sector. This switching was achieved through the actions of banks, investors and developers. A number of other sources of investment complemented the flow of petrodollars into the economies of the West. These included interest payments from Third World nations of their debts, investment from the growing economies of the Middle and Far East and investments from growing European pension funds (Harvey 1989b; Knox 1992a; 1993: 4–7). The symbolic (and very probably actual) end of this period of urban development was the bankruptcy in 1992 of Olympia and York, the developers of Canary Wharf in London's Docklands. The over-investment of the 1980s caused the collapse of the property market in the early 1990s.

While the pace of development was fuelled by the influx of investment, its shape, character and imprint on urban landscapes and urban form were determined by other factors. The development industry has undergone a considerable transformation in recent years becoming both more concentrated and centralised into a smaller number of very large firms. Smaller firms have been eliminated to a large extent through competition and merger from larger firms. The average size of firm in the development industry has increased (Knox 1993: 5). The consequence of this for the urban landscape has been that these larger firms are able to undertake developments on a larger scale than were previously the case. This is evident in a number of new megastructures dotting the urban landscape (Crilley 1993; Knox 1993: 5–6). Increasingly in certain 'hot' sites, usually near the centres of world cities, property development has become a less local activity (Strassman 1988; Leyshon *et al*. 1990; Knox 1993: 6–7). The development of these sites has become increasingly bound up with the workings of the international economy. These megastructures have been developed mainly by large multinational developers, such as Olympia and York, who have relied on heavy borrowing from banks only too willing to lend to the booming property sector (Knox 1993; Sudjic 1993: 34). Formerly run-down regions have been favourite sites for such developments. Examples have included Battery Park in New York, California Plaza in Los Angeles and Docklands in the East End of London. It has been projects of this scale and staggering architectural style that have demonstrated the power of the large developer in reshaping or reorienting the landscapes of certain cities.

> While Canary Wharf was shaped by unabashed commercial opportunism, its closest contemporaries, the World Financial Center in New York, and California Plaza in Los Angeles, paid at least lip service to the strategic planning guidelines set for them by city governments. But planner or not, all these schemes served to demonstrate that it is the property developer, not the planner or the architect, who is principally responsible for the current incarnation of the western city. Large scale speculative developments – offices, shopping centres, hotels and luxury housing – shape the fabric of the present day city, not public housing and civic buildings. The developer, or more likely the institutions that fund his projects, pays the price for the land on which development will take place, determining far more rigidly than any zoning ordinance the range of activities that it may be used for. He [*sic*] chooses the architect and he sets the budget.
>
> (Sudjic 1993: 34–5)

In the same way that other sectors of economic production have been able to become more flexible with the incorporation of computer technology and flexible labour practices into the production process, so has the production of the built environment (Knox 1991; 1993). This has mirrored the shift in production away from large-scale production dependent on economies of scale, towards flexible production aimed at the exploitation of highly profitable niches within the market, and generating profits through the range or scope of products available (Harvey 1989a). This emphasis on individuality and product differentiation is evident in new commercial, retail, leisure, industrial and residential landscapes (Knox 1993: 7–9; Graham and Marvin 1996).

This emphasis on product differentiation has had a reciprocally reinforcing impact on the culture of the architectural profession. This profession has become increasingly divorced from the social ideals which guided its earlier development (see Jencks 1984; Harvey 1989a; Knox 1993). In previous periods, for example in the early post-war period, architecture, along with practices such as town planning, was suffused with social idealism. In the UK, for example, the design of the urban environment was seen as being important in both reflecting and helping to mould an egalitarian, democratic society, fit to accommodate the returning 'heroes' of war. It stood alongside the development of social welfare programmes of the welfare state and the National Health Service in the UK as one of the central canons of post-war society. Earlier architecture had regarded itself as a practice fashioning social utopias rather than just physical structures. This was the case with both the Swiss architect Le Corbusier's *'unit d'habitation'* and the earlier garden city movement.

The loss of architecture's social vision has coincided with its being co-opted by large institutional investors and speculative developers. This emphasis for the architectural profession has shifted to one of satisfying the demands of their clients, often engaged for a single project. This is typically achieved through decorative design and stylistic distinction, the massive scale of developments or the employment of a publicly known 'superstar' architect whose presence on a project will attract a great deal of media attention (Crilley 1993). Architecture, together with the architects responsible, has increasingly become a form of corporate advertising, and despite a few remaining radicals, has retreated from what it saw as its earlier social purpose. The consequence of this sea-change in the architectural profession has been that the urban landscape is designed in (artful) fragments and becomes littered with a number of 'spectacular', 'imageable' or 'scenographic' enclaves which are largely divorced from their immediate urban or social contexts (Harvey 1989a; 1989b; Crilley 1993; Knox 1993).

The regulation of urbanisation

The prime regulatory mechanism of urbanisation is the planning system. However, the culture of planning, very much like that of architecture, has also become increasingly divorced from its social origins. This has been so for a number of reasons, one being the failure of post-war planning to deliver the social utopias it had promised. This failure was best exemplified by the failure of the high-rise residential towerblock in a number of British cities. These have become some of the most blighted residential environments and many face demolition. During the 1980s interventionist planning became regarded by the New Right as an impediment to the successful, free operation of the market. The New Right sought to reduce the influence of planning over the operation of the market as part of a sweeping reform of the public sector under the banner of 'freeing the market'. The planning system, through agencies such as the Urban Development Corporations in the UK, became a mechanism for the facilitation, rather than the regulation, of urban and economic development.

This shift away from the regulatory role of planning helped usher in the fragmentary development of the urban environment. The planning system was able to conceive of the development of the city in its totality, and to situate developments within their wider urban context. Developers who became increasingly influential agents of change in British cities during the 1980s, by contrast, have no need to look beyond the boundaries of their own development projects. They are primarily concerned only with the planning of their own 'fragments'. Planning in the 1980s and 1990s had become little more than the buttress of this fragmentary approach to urban development.

The consumption of the urban environment

None of the changes in the production and regulation of the built environment would have been likely to occur were it not for a concomitant change in the patterns and practices of the consumption of the urban environment. Put simply, new urban landscapes would not have appeared if no one was going to buy them. The relationship between supply and demand in this case is a complex one. It is difficult to say which had the major determinate influence; however, it is clear that they were reciprocally reinforcing.

It is true to say that contemporary urban culture is very consumption oriented. For those who are able to afford it, it acts as an important marker of status, distinction and identity. Consumption has always had this function to some extent; however, the patterns of consumption that arose during the 1980s were distinctly different from those of earlier. The emphasis shifted towards notions of exclusivity, style and distinctiveness. Consumption at the top end of the market shifted away from the consumption of mass produced goods, characteristic of the thirty years following the Second World War. Consumption patterns in the 1980s fragmented into a series of niches determined by lifestyle or cultural preference.

Understanding the reasons behind this fragmentation of consumption patterns requires an appreciation of the psychological value of consumption. Consumption is a fundamental part of both individual and social identity construction. This applies equally to the consumption of places as well as commodities or ideas. Consequently the geographies of places of consumption (residential, commercial, retail and leisure) and the consumption of places is an important aspect of urban geography.

The new emphasis on consumption with urban culture derives from two phenomena, one social and one economic. First, the post-war 'baby-boomer' generation (born between 1945 and 1955) was characterised by an accent on the exploration of self-awareness. During the late 1960s this was achieved through the widespread adoption of a number of counter-cultural movements. However, the failure of these movements to deliver such self-awareness and freedom led to this urge being satisfied through the pursuit of individualised patterns of consumption (Knox 1991). This has become known as 'from hippie to yuppie' (Ley 1989). Second, the polarisation of economic opportunity and income, a pervasive economic characteristic noted elsewhere, created with them a number of jobs at the upper end of the spectrum that did not carry with them any inherent social status. This missing social status had to be constructed. Again consumption was the mechanism through which this was achieved (McCracken 1988; Knox 1991: 183–6; 1993: 19–25; Jackson and Holbrook 1995). As a result of this demand for status through consumption a number of key urban landscapes were reconfigured around conspicuous consumption. These were sites where consumption was conspicuous and conspicuous sites which were consumed.

Recent urban change

This section broadly surveys some recent changes in the forms and landscapes of cities since the late 1970s and early 1980s. It will examine in turn changes in the city centre, the inner-city, the suburbs and will conclude with a look at the emergence, or otherwise, of post-suburban developments in the UK, drawing comparisions with international developments.

The city centre and inner areas

The centres of British and North American cities had, by the early 1980s, become much maligned places. Blighted by a legacy of poor post-war planning and design, they were typified by poor physical environments, pedestrian unfriendly traffic systems, downgrading retail environments and economies dominated by offices that relegated the cities to cultural deserts after the early evening. Most importantly they were largely devoid of any economic mechanisms which could make positive impacts on the local economy.

Since then, however, city centres have been the focus of an astounding degree of development. The 1980s have shown a widespread and comprehensive 're-imagination' of city centres (Bianchini and Schwengal 1991). This process has involved a combination of physical enhancement and cultural animation processes which should be viewed in parallel to the transformation of the images of cities which is described in chapter 7. This re-imagination of the city centre is manifest in a number of landscape elements that have emerged in city centres during the 1980s, and urban policies that have aimed to animate these spaces through the promotion of an enhanced public or street cultural life.

Spectacular and flagship developments

Without doubt the most prominent landscape element to emerge in city centres during the 1980s was the spectacular or flagship development.

These developments are of many kinds; however, what they have in common is their large scale and the emphasis on the importance of eye-catching, decorative, spectacular or innovative, typically post-modern, architecture. Flagship developments include office developments such as Canada Tower at Canary Wharf in London's Docklands, Convention Centres such as the International Convention Centre in Birmingham and the Scottish Exhibition and Conference Centre in Glasgow, Sports Stadia such as the Riverside Stadium on Teesside, the new home of Middlesbrough Football Club, the National Indoor Arena and the Don Valley Stadium near Sheffield, industrial parks such as the Dean Clough Industrial Park in Halifax, museums such as the National Museum of Film and Television in Bradford and a whole range of temporary events such as sports tournaments, cultural, arts and garden festivals. These developments and events have frequently been initiated through some form of public–private partnership between local government or their agents and the private sector. Agencies such as the European Union have often been targeted as potential sources of funding.

International examples of this include the redevelopment of Battery Park City in New York, La Defense project in Paris and various spectacular office projects in the previously neglected centres of American cities such as Los Angeles, Houston, Cleveland and Atlanta. The reunification of Berlin since 1989 offers an interesting case of this process. The city's political and physical division, expressed by the Berlin Wall running through its heart, had inhibited and distorted the development of its central core. However, since the reunification of Germany the Berlin Wall has been dismantled, opening up a vast swath of derelict land through the city for development. The result has been a scramble by speculative developers to regenerate the city's heart spectacularly.

Such developments have both a tangible economic function as well as a less tangible, but no less important, symbolic function (Bianchini *et al.* 1992). These developments were primarily intended to act as catalysts, kick-starting the regeneration of the local urban environment and economy. Physically such developments are able to bring derelict land back into use and upgrade or enhance existing land-uses. They also act as economic 'magnets' which are intended to attract people, spending and jobs, according to their type, in theory stimulating the wider urban economy (Harvey 1989b). They frequently act as stimulants to other economic development strategies by the local authority. It is from such 'ripple' effects that flagships derive their nickname. They not only attract or stimulate a 'flotilla' of smaller developments within cities but also may promote the development of policies or strategies that aim to spread the effects of development across the city (Bianchini *et al.*1992).

Symbolically these flagship developments can act as central icons in the apparent transformation of a city's fortunes, image and identity (Bianchini *et al.* 1992: 250; Crilley 1993; Hubbard 1996). In the post-industrial economy, image is a vital component of economic regeneration (Watson 1991). Flagship developments can act both as important symbols of the rejuvenation of a city and icons around which new images for that city might be constructed (Hubbard 1996). The 'scenographic' or 'cover shot' qualities of flagship developments are important in this respect (Crilley 1993; Knox 1993). The involvement of an enthusiastic local media, politicians and councillors is often vital in this effort (Thomas 1994). The hype that surrounds flagship developments is as vital a part of their function as any economic rationale behind their development.

Flagship developments of the types discussed here first appeared in the USA. The first example was probably the Charles Center, a mixed-use development in Baltimore, which was built in the late 1950s. This was followed by a number of other spectacular developments around the city's Inner Harbour area (Bianchini *et al.* 1992: 246). The models of development pursued in Baltimore have proved hugely influential on those developed in the UK during the 1980s. The influence is apparent, for example, in the Broad Street Redevelopment Area in central Birmingham. This originally contained plans for 'Brindleyplace', a festival retail development based directly on Baltimore's Harbourplace development.

Flagship developments have been frequently sited within equally eye-catching 'packaged-landscapes' (Knox 1992a), which are designed in conjunction with flagship projects in order to complement the functions of the flagships. These landscapes include various permutations of architectural renovation, heritage themes, waterside developments, specially commissioned public art programmes and newly designed or renovated public squares.

The British fashion for flagship development was primarily initiated by the change in the political climate which forced local authorities to become more entrepreneurial and fight to lure capital investment and jobs into their cities, rather than to expect financial help from central government. The flagship model of development was seen as a successful method of achieving these ends. Flagship redevelopment also reflected the post-modern conception of space inherent in the culture of much contemporary architectural practice and the planning profession. Flagships can be regarded very much as 'artful fragments' or 'scenographic enclaves' (Crilley 1993). They represent spaces which are shaped for aesthetic rather than social ends. There is little thought for how they might fit into the wider city economically, socially or culturally. Indeed their failure to integrate into their surrounding social fabric (discussed in Chapter 8) gave rise to a number of descriptions such as 'yacht havens in a sea of despair' (Hudson 1989) which reflected the juxtaposition of conspicuous displays of wealth in the developments themselves and the, often severe, social and economic deprivation which surrounded them.

Festival retailing

City centres have traditionally been important retail foci. Although retailing has displayed a trend towards decentralisation during the 1980s, certain parts of the city centre retail landscape underwent considerable rejuvenation during the 1980s through the spread of festival shopping environments. This trend was closely associated to the spectacular redevelopment of the centres of British cities. Indeed many festival shopping developments can be regarded as flagships in their own right. Again this followed earlier developments in the centres of North American cities.

Festival shopping can basically be regarded as up-scale shopping within themed environments, or as the product of the collision of two of the most important trends in urban culture during the 1980s, the physical revalourisation of previously redundant space and the construction of identity through consumption. Festival retailing can be thought of as the result of the shift towards consumption which highlighted the importance of exclusivity in the construction of identity. The retailing boom of the 1980s was facilitated partly by the explosion of disposable income and the availability of personal credit among the young, affluent and middle class during this period.

The cultural shifts within which consumption was implicated made it far more than merely a functional fulfilment of need but a significant leisure activity in its own right. It became important, therefore, that its environments strove to convey an air of exclusivity. This was achieved through the use of ornate, post-modern design, often employing historical quotation and pastiche, through the theming of malls, or through claiming the cultural capital using buildings with historical associations and/or waterside locations. These developments often included restaurants and wine bars as well as shops. Festival retailing is a practice where the consumption of space and time is of equal cachet as the consumption of designer goods (Harvey 1989a).

Landscapes of heritage and nostalgia

The reclamation, renovation, re-use or reconstruction of past urban landscapes has become an almost ubiquitous aspect of the contemporary urban scene. These landscapes of heritage and nostalgia take on many forms. They include museums of urban and industrial history, the kitsch historical adornments of many packaged landscapes and new developments such as industrial hardware recycled as forms of street furniture, the renovation of old buildings or districts providing commercial, industrial, recreational or residential property and the do-it-yourself renovations of inner-city gentrifiers. Similarly the scale of these landscapes ranges from the minute adornments of historical signs or designs up to the thorough renovation of whole districts, in or near city centres.

The popularity of heritage and nostalgic elements within new urban landscapes in the UK can be understood within the context of a national culture which, while having had a strong conservative and heritage tradition, has recently shown a renewed interest in nostalgic, idealised views of the past. This has been manifest in an explosion of heritage museums, nostalgic television programmes and films, writing, advertisements and other media. Many commentators have interpreted this as a retreat from the dreadful conditions that prevailed in modern Britain since the mid-1970s (Hewison 1987). However, an alternative and more positive interpretation of the heritage impulse has argued that this interpretation is overly elitist and dismisses the validity of popular conceptions of history. Critics of the earlier interpretation have argued that the heritage industry offers a more democratic and accessible experience of history (Samuel 1994).

In an international context the incorporation of past landscapes into the contemporary urban scene also fits with the post-modern turn in architecture and urban design. Namely it is able to satisfy the post-modern yearning for eclecticism, the vernacular, regional distinctiveness and decoration through historical quotation. Heritage, or rather the very obvious possession of a past by an urban space, expressed either through a famous past or through prestigious historical architecture has proved a valuable commodity of great appeal to consumers of urban space in the 1980s and 1990s.

Cultural animation

While the physical changes described above might provide the template for the re-imagination of a 'new' city, physical change is not all that is required to regenerate city centres. It became obvious during the 1980s that public culture in many city centres was very impoverished. The centres of large cities in the UK and North America provided little more than spaces for working and shopping. An active public cultural life was regarded as a necessity for successful regeneration of urban space. By contrast many European cities were known for their vibrant, lively public culture. Where spaces for such public culture existed in British and North American cities, by the early 1980s they were downgrading, a legacy of their generally poor planning, or they had been appropriated, in popular perception at least, by anti-social behaviour. The rejuvenation of public urban culture in these cities in the 1980s was achieved through the physical improvement of city centre public space and a range of local authority policies which aimed to improve the safety and accessibility of city centres and to encourage open-air, free events. Policies have included street lighting, pedestrianisation, relaxed licensing hours, the introduction of closed circuit television, traffic calming and improved public transport (Bianchini and Schwengal 1991).

The provision of new or redesigned public squares, often complete with extensive programmes of specially commissioned public art and entertainment such as street theatre and musical displays, has been a common response by city councils. Birmingham has replanned two major civic squares in its centre during the 1990s. Both had large budgets for the provision of public art. Similarly in Sheffield, Tudor Square has been transformed into the setting for a number of free public events.

The shifts in design and policy outlined above reflect a recognition of the importance of the 'night-time' economy, both to the urban economy and to the enhancement of the cultural life of the city (Montgomery 1994). This was a very deliberate attempt to reorient urban culture more along European lines than has previously been the case.

The inner-city: a sea of despair?

In debates about the form and character of the post-modern city, the inner-city is conventionally painted as the 'sea of despair'. It is portrayed as a zone abandoned by both formal economic mechanisms and conventional forms of social control and regulation. These, it is argued, are replaced by forms of twilight or informal economy and social control based upon violence or threat. These images first emerged in coverage of social and racial disturbances in areas such as South Central Los Angeles. These images formed the staple coverage of a whole variety of media, both fictional and non-fictional (including some academic coverage). Such nightmare images of the inner-city have, since the early 1980s, begun to appear in accounts of the inner areas of a number of British cities. The inner-city has come to represent the dark side of the post-modern city, sharply juxtaposed with the spectacular redevelopment of the city centres. However, it is important not to let the moral panics stirred up by such representations cloud the reality of life in inner-cities.

The inner-city possesses both a formal geographical identity and a metaphorical identity. Geographically the inner-city is that area between the city centre and the suburbs, traditionally referred to in the geographical models of Burgess and Hoyt as the 'twilight zone' or the 'zone in transition'. It was an area of high-density housing, largely terraced, which dated back to the late nineteenth century mixed in with local authority housing redevelopments. It grew up around nearby heavy industry before it was closed down or relocated to the suburbs, creating another set of problems for the inner-city. The inner-city was primarily associated with working-class and immigrant populations. However, this strictly geographical definition has become somewhat surpassed by its metaphorical identity. The rigidly delimited zoning and internal homogeneity of the urban form of the industrial city has been dissolved to some extent recently. As a result of this the term 'inner-city' has now come to refer less to a specific geographical location and more to a set of social, economic and environmental problems perceived (often incorrectly) as being typical of the inner-city. Such problems, however, are as likely to occur on municipal housing estates on the edges of cities as they are in the geographically inner areas of cities.

The inner-city has always been regarded as the lower-class environment of the city. What is distinct about recent commentary, however, is the equation of the inner-city with the existence of an urban underclass. The inner-city has become regarded, not as the poor relation of the city, but as a member of a different family altogether. This echoes wider debates about the polarisation of the economy and the severing of the links between groups at the lower ends of the spectrum and those at the top. The inner-city has become the imagined nightmare environment of the disenfranchised and the excluded.

Such moral panics of urban spatial and social forms have begun to be transferred into new urban forms such as recently designed suburban settlements for the affluent middle classes in cities like Washington, DC and Los Angeles. These settlements, such as Park La Brea and CityWalk in Los Angeles, display a desire among the middle classes for surveillance of their property, protection and security (Beckett 1994). This 'citadel' or 'paranoid' architecture is in response to a perception of the poor as 'other', a threat (Soja 1996: 204–11). These panics are whipped up by selective coverage of these areas in the media, coverage which focuses only on the newsworthy aspects of these areas, crime, violence, drugs, gangs and apparent social breakdown.

This perception of the inner-city is reflected in the geographies of service provision and abandonment. The inner-city constantly emerges as the area most qualitatively and quantitatively under-provided with basic health, welfare, education and financial services. Banks, for example, appear increasingly to be adopting a policy of 'cherry picking' their customers on the basis of income. This inevitably has a geographical expression in terms of access to financial services. The policy is expressed in two ways. First, the geography of financial service provision is changing, with banks closing outlets in inner city areas and shifting provision towards areas with more affluent social profiles (Leyshon and Thrift 1994). Second, an increasing proportion of bank business is transacted electronically. In 1988 Midland Bank set up First Direct, the UK's first 'electronic bank', which uses electronic telecommunication systems to deliver all of its services and has no physical branches on the high street, only one single central office. The bank's customers are specifically chosen on the basis of their incomes, focusing on those of an income above a certain level (Graham and Marvin 1996: 149–50). The distribution of these customers, therefore, reflects that of the traditional banking outlets, concentrating in areas of affluence over areas of poverty. This process is referred to as 'financial exclusion' (Leyshon and Thrift 1994).

Such attitudes towards the inner-city are not new. In many ways the contemporary representation of the inner-city reflects very closely those of middle-class Victorians in both Britain and North America. The rapid and largely unregulated growth of the Victorian city generated a series of moral panics centred on fears of crime, social disorder and disease (Coleman 1973; Mayne 1993). These were a staple of satirists, journalists and social reformers at the time.

Returning to the more strictly geographical identity of the inner-city, it is certainly apparent that examples of the 'seas of despair' discussed above do exist. However, this image should be qualified. The inner-city is a far more multifaceted entity than this image would suggest. First, simply to regard all inner-cities as constituting seas of despair is a vast oversimplification. It hides far more than it reveals. Second, it is worthwhile asking if the conditions existing in inner-cities are worsening. Are inner-cities progressing along a common trajectory towards becoming seas of despair? While there is evidence that in general the problems typically associated with the inner-city are worsening, to label this trend universal would be to oversimplify a very complex series of trajectories. This label ignores the fact that while the economic factors affecting conditions in inner-city areas might be general, they are mediated locally. The outcome of this interaction of the general and the local is a complex mosaic of local difference. The local economic, political, social and cultural conditions of inner-city areas are as crucial to highlight as the general economic processes with which they interact. Consequently, for every inner-city area or peripheral council estate apparently in the grip of economic decay and social dislocation and breakdown, there are many others, where, despite conditions being worse relative to other areas in the city, the reality does not fit the prevailing media representation. Even within areas notorious for economic, social and environmental problems, such conditions are rarely universal. Such areas, despite media representations to the contrary are internally heterogeneous. Conditions within them are far from universally dreadful.

Media portrayals of the inner areas of cities are selective and partial. That which is deemed newsworthy is by definition exceptional. The ordinary and the everyday simply do not get reported in newspapers and on television. There is a danger that in examining only the exceptional events that media representations are mistaken for reality. This stigmatises inner-city areas and peripheral housing estates. It may be that conditions and opportunities in urban areas in general are becoming more polarised but at a local level this process crystallises out as a mosaic of local complexity which belies such simplistic images as the 'sea of despair'. Inner-city areas should be recognised as complex; however, providing a detailed guide to such complexity is beyond this short section. Rather, it has tried to provide a framework within which the character of inner-city areas can be appreciated.

Gentrification of the inner-city

Gentrification is a term that has come to refer to the movement of affluent, usually young, middle-class residents into run-down inner-city areas. The effect is that these areas become socially, economically and environmentally up-graded. Gentrification is a process that has generated ongoing geographical debate and dispute since the early 1980s.

It is generally argued that these middle-class residents, drawn by the appeal of the 'buzz' of inner-city living, are attracted to the cultural capital attached to distinctive architecture, often available at very low prices because of its physical dilapidation (Fincher 1992). Such a view, however, disguises considerable variation in the nature, extent and impacts of gentrification. While gentrification is an established and extensive practice in many North American, Australian and some European cities, in the UK, with the exception of London, it is less apparent. Similarly there is a great variation in the groups involved in the gentrification of the inner-city. As well as the young, middle-class professionals, artists seeking cheap extensive work space (Zukin 1988) and women seeking access to city-centre employment opportunities and good public transport have been identified as gentrifiers (D. Rose 1989; Fincher 1992: 107). Local authorities have promoted a number of activities that support gentrification, such as the refurbishment of old industrial space to provide studio space for artists and cultural industries, as a way of economically revitalising and diversifying inner-city areas. One well-appreciated aspect of gentrification is the potential impact it has on existing working-class residents. These impacts have included social disharmony as new groups enter community space and the displacement of working-class groups as house prices rise in the wake of gentrification. These problems have been very apparent in the Urban Development Corporation-led gentrification of London's Docklands (G. Rose 1992) and the gentrification of the Lower East Side in New York, lead by the real-estate industry and encouraged by the City (Reid and Smith 1993). Clearly the extent, diversity, institutional linkages and impacts surrounding gentrification should not be overlooked (N. Smith and Williams 1986; Hamnett 1991).

The suburbs

The suburbs can be defined as the outer areas of a city which are linked to the city by their lying within the commuter zone of an urban area. They usually form a continuous built-up area. 'Suburbs' usually refers to the predominantly residential landscapes built up around the urban core as a city has expanded outwards. Suburbanisation of the residential population of cities is essentially a twentieth-century process closely associated with development of transport technology: the trolley-bus, the train and the car respectively. The suburb has become the landscape of the middle class and the skilled working class internationally.

Despite being burdened with an image that suggests monotonous regularity, the suburban areas of cities, or indeed of any single city, display a great variety. This variety in large part derives from the different periods during which suburbs have developed (Whitehand 1994), the markets for which they developed (Knox 1991; 1992a; 1993), their physical settings, the architects and developers engaged and the local planning framework and operation of the system (Whitehand 1990). Again in contrast to their image of unchanging regularity, the suburbs are dynamic landscapes which demonstrate complex processes of change (Moudon 1992; Whitehand *et al.* 1992).

Suburban development and change

Suburbs have tended, over the course of the twentieth century, to expand outwards through addition. A number of phases of growth can usually be recognised within cities which relate to a number of social, economic and technological factors. These have given rise to a series of distinctive landscapes. In the UK these phases include Victorian, inter-war, post-war and recent development. Each is distinct in the landscape it produces.

Although these phases of development have been successively outwards, they have not been uniformly smooth. The pace and style of suburban development is related very closely to booms and slumps in house building cycles. Slumps in house-building, related, for example, to credit availability, interest rates on borrowing and demand for housing, generally cause land values to fall. This has a number of effects on the amount of house building occurring, the size of houses and plots and other types of building and land-use included in suburban areas. During

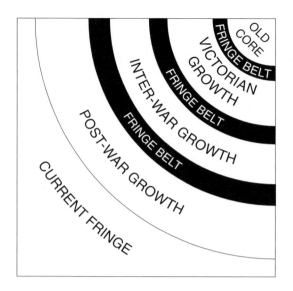

Figure 6.1 *A model of suburban development*
Source: Whitehand (1994: 11) by permission of Geographical Association

slumps it becomes feasible to build more houses at lower densities on larger plots of land and to build more non-residential buildings such as institutional or commercial buildings and to include extensive land-uses such as sports pitches and playing fields. The characteristic landscapes formed during building slumps are referred to as fringe-belts. During building booms, outward development tends to be more rapid, housing densities higher and non-residential land-uses fewer (Whitehand 1994).

Other important influences on the form of suburban development are innovations in building and transport technologies and the pace at which these are adopted. Evidence has shown that there has been a particularly close association between the adoption of various innovations in transport and bursts of house-building activity, again with each boom producing a characteristic landscape.

A model of suburban development has been proposed which includes each of these factors of change: additive growth, fluctuations in building cycles and building and transport innovations (Figure 6.1).

Recent suburban change

In contrast to the more dynamic city centres the suburbs of British cities have recently been subject to changes that have been more piecemeal than epochal. Once built, suburban residential zones are far from static, and display change throughout their lives. The suburban landscapes of Britain have changed predominantly through processes of addition, replacement, refurbishment, conversion and conservation since the mid-1970s. One of the most important changes in the British suburban landscape has been the conversion of single unit dwellings into multi-unit dwellings, such as flats and the infilling of vacant land in large plots, usually the gardens of large detached houses. Although these have been widespread processes across the UK they have demonstrated some regional differences, with the South East region under greatest pressure to increase dwelling densities in its suburban areas (Whitehand 1994). This process is in sharp contrast to the rapid and extensive outward expansion of suburban areas in the first half of the twentieth century which was fuelled by, and in turn fuelled, the rapid spread of the motor-car.

The redevelopment of Britain's existing suburbs and the development of new residential areas is likely to become a major planning issue into the early part of the twenty-first century. The Department of the Environment forecast in 1992 that an additional 4.4 million households will need to find homes in England between 1991 and 2001 (Breheeny and Hall 1996). This raises the obvious question: where will these homes go? Not only is it unlikely that the existing housing stock will adequately provide for this increased demand in terms of sheer numbers, but also it is probable that the existing housing stock will not provide houses of an adequate type for this new demand. The outward expansion of British suburbia proceeded on the assumption of an 'ideal' household which consisted of a nuclear family. However, such an ideal type is increasingly not the case in the UK, which has seen an increase in alternative types of households. These have included single-headed households, extended families, couples delaying marriage and children, and an increasing number of people choosing to live alone. This demographic shift has led to a demand for several different types of housing. It is likely that new dwellings of different types will have to be built to cater for this new demand and that much of this new building will be focused on smaller urban settlements. The possibility of building new towns or settlements on greenfield sites has also received renewed interest.

The outward expansion of towns and cities has been restricted through the introduction of restrictive planning policies such as the 'green-belt' policy in 1938. The green-belt policy, and its variations, has been designed to ensure the protection of vacant green land between cities. The policy was intended to prevent the unchecked urban sprawl swallowing up the countryside between cities. Despite coming under considerable pressure since the late 1970s it remains a powerful impediment to urban growth.

The meanings of suburbia

As well as possessing a very distinctive physical identity, British suburbia possesses an equally distinctive cultural 'identity'. This is evidence of the fact that it is the repository of a number of important cultural values. English culture has long demonstrated a strong anti-urban streak which can be traced back to the middle of the nineteenth century. By contrast, it has also displayed a powerful veneration of an idealised 'chocolate-box' vision of rural England. The suburbs can be read as a response to the need to live near the centre of cities while wishing to attain an essence of rurality. Despite being essentially part of the urban settlement, suburban areas have long been imbued with associations of the rural, through their physical design, which has been centred on the garden, their naming and their promotion and representation (Gold and Gold 1994) (Plate 6.1).

The continued manipulation of this image has ensured that suburban areas have become imbued with values long attached by the English middle classes to the rural idyll: tranquillity, peace, community, safety, an unhurried pace of life and domesticity (Eyles 1987). This identity has become the focus of sharp criticism by feminist geographers.

Plate 6.1 *Early building society advertisement for suburban homes*
Source: Gold and Ward (1994: 83) by permission of Abbey National Building Society

Geographical perspectives on the suburbs

The suburbs have been subject to a number of different interpretations by geographers adopting contrasting perspectives. While they are not competing interpretations, because they are concerned with different aspects, they emphasise the range of economic, social, psychological and cultural perspectives geographers have adopted to study the urban, and the suburban.

Marxist geographers have interpreted the massive suburban expansion of the cities of the USA, Canada, Australia and the UK in the inter-war and post-war years, as an attempt to resolve a potential crisis of over-accumulation in capitalist economies. Overaccumulation in the primary sector (which includes industry) was resolved, it is argued, by capital 'switching', through the economic policies of central governments and the actions of lending banks, to the secondary sector, which includes property development. This, they argue, created a boom in property development fuelling suburban expansion.

By contrast both humanist and feminist geographers have developed critiques of the landscapes and cultural assumptions which have underpinned suburban development. Humanists have highlighted the lack of originality in the design of suburban landscapes, their monotony and lack of regional distinctiveness. This, they have argued, has failed to provide these landscapes with any 'sense of place', a vital psychological component of human feelings of security and belonging. This perspective has been criticised by sociologists and geographers, who argue that this is an elitist perspective which ignores a variety of attachments to ordinary landscapes that belies their uniform physical appearance.

Feminist geographers have attacked the sexist stereotyping inherent in the production and promotion of suburban landscapes. The design and depiction of the suburban house and its landscape promote notions of domesticity, consumption and recreation. This, they argue, hides the fact that these landscapes are largely maintained by women, who, because of male dominance of car use and poor public transport provision, are daily trapped in these areas. The image of suburbia which hides the work required to maintain the ideal of this landscape is part of the general failure to regard housework as 'real' work. The feminist critique provides a valuable corrective to the ideal of suburban life represented in Plate 6.1.

Post-suburban developments in the UK

Observers of the landscapes around the edges of major cities in the USA have begun to recognise the emergence of a new form of residential landscape that has been termed 'post-suburban'. These developments include extensive, private, master-planned developments that are radically different from the traditional residential suburb. These developments are aimed at the affluent consumer. Individual dwellings are typically large and situated in extensive, well-landscaped grounds. The design of individual dwellings and the landscaping of whole developments are closely based on ideas of tradition and rurality borrowed heavily from idealised Anglophile notions of the rural realm (Knox 1992a). These developments are marketed and named accordingly, in formerly undeveloped plots, and come provided with a wide range of amenities, normally found in towns, which are planned into them. These include a town centre, public squares, police and fire stations, libraries, theatres, meeting halls, schools and post offices. All of these are designed to fit into the overall motif of the development. They surround, or are found in close proximity to, significant concentrations of offices, shops and other formerly 'down-town' commercial functions which have decentralised, driven by high land costs, central city disorder and freed by telecommunications developments. These, rather than being an extension to the urban area, which is the case with conventional suburbs, can be regarded as an alternative to the city; they have come to be known as 'edge-cities' or 'stealth-cities', a reference to the fact that they frequently straddle a number of administrative districts and, therefore, fail to crop up on official statistics and census returns (Garreau 1991; Knox 1992a; 1992b; 1993).

What is distinctly different about these developments is their origins in the private sector, their development which was fuelled entirely by the development boom of the 1980s and the control exerted over their design by a single developer. In the USA they have rapidly become significant elements of the urban geography of North American cities. The proliferation of these developments around major cities has somewhat dissolved their traditional form and created an urban form which has been referred to as a 'galactic metropolis' (Lewis 1983; Knox 1993).

The immediate fringes of British cities and the areas beyond have been the locations of considerable development, with, for example, the decentralisation of retail and industrial land-uses, the development of business and science parks and the continued pressure of the outward spread of housing. The most developed example of what should be called the 'spread-city' in the UK, rather than the edge-city, is London and the surrounding South East region. A considerable amount of the economic activity of the South East region has developed along axes from London. These have included the three motorway axes, the M3 axis from London to Southampton which includes Guildford, Basingstoke and

Southampton, the M4 axis from Heathrow Airport through Slough and Reading towards Swindon and extending to Cheltenham and Bristol (this area contains around 60 per cent of the new high-technology firms established since the mid-1970s) and the research and development and aerospace oriented axis along the M11 through Hertford towards the Cambridge Science Park established by Trinity College. The restructuring of these areas is closely linked to the geography of government research centres which include Aldermaston (weapons research), Farnborough (aircraft) and Harwell (nuclear). These areas, as well as a number of historic towns in the North, such as York and Lancaster, have developed economies based around technologically advanced methods of production and social profiles showing high numbers of the emergent, affluent 'service class' (employers, managers and professionals). Such developments have had a significant impact on the British urban scene. However, they have contributed to a general decentralisation of British urban form, rather than the dissolution that geographers have argued is the case in the USA. Their regional concentration in the South East and limited impact on the overall form of other British cities appear to mark these developments out as an exception within British cities rather than the harbinger of the age of the British edge-city. The rediscovery of the city centre as a focus for urban cultural life and the predications that the actual amount of telecommuting in the future is never likely to match the potential the technology offers would support the notion that the traditional focus of British urban form has not yet been superseded by a post-suburban equivalent. The differences between the urban systems of the USA and Europe suggest that, while the edges of European cities are increasingly important foci for the study of urban development, simply importing theoretical perspectives derived from the study of US post-suburban development is unlikely to prove enlightening. The differences between Europe and North America and the differences within European urban systems need to be appreciated to understand new urban forms appearing around the edges of existing cities (Keil 1994).

Post-modern urbanisation: a blue-print for urban change?

This chapter has been concerned with an apparent transformation ('post-modernisation') of the processes of urbanisation and some of the resultant forms. The products of this post-modernisation are very apparent in parts of North America, especially in the hi-tech-oriented world cities such as Los Angeles and Washington, DC. Even a cursory review of the British urban landscape, for example, reveals that such post-modern urban landscapes have not yet emerged on such a large scale. London is the best example of a city whose centre has been transformed along the lines seen in other world cities around the world. This is closely related to its position as a, possibly *the*, pre-eminent world financial centre, while its suburbs and surrounding regions are the best exemplars of the economic and social restructuring associated with the rise of hi-tech industry. This is related to its wealth of research facilities, the presence of an appropriate workforce and the pleasant suburban environments for them to live in. However, these point to the fact that these are largely the exception rather than the norm within the British urban system. Certainly, examples of post-modern urban form can be found elsewhere: the service-oriented yet 'historic' settlements of Lancaster and York, traces of post-modernism within otherwise largely 'modern' city forms, and the spectacular redevelopments of city centres surrounded by areas of social and economic deprivation. However, it would be inaccurate to declare the British urban landscape unequivocally post-modern; large areas of many British urban settlements have been largely unaffected by the apparently epochal forces of post-modernism. Likewise, the partial, rather than total, transformation of urban form is the case for most cities in Europe and Australia. Post-modern urbanisation, like most other facets of urbanisation, is emerging as a complex series of trajectories mediated locally rather than a single, simple, universal trajectory of development.

There are three reasons for this state of partial transformation (or rather one reason expressed three different ways). First, the processes of post-modern urbanisation are general processes; however, they are mediated by local conditions. Their outcomes depend on the mediating effects of local social, economic and cultural conditions and the effects of individual actors in particular situations. The existence of general processes of post-modern urbanisation does not automatically and unproblematically translate into the production of post-modern cities. Second, cities are not passive receptors of change. While general processes of urbanisation might try to make their imprint on urban landscapes they meet with resistance and are tied by the legacy of existing urban landscapes. Cities are not easily and infinitely flexible and they exert a considerable influence over the operation of general processes of urbanisation. Again the operation of general processes of post-modern urbanisation does not necessarily translate into urban transformation. Finally, urban hierarchies and individual cities are internally heterogeneous. General processes of urbanisation are refracted through these patterns of internal differentiation. Consequently while some cities and some areas of cities are affected by the processes of post-modern urbanisation, others are not. In conclusion then, the notion that we are witnessing a general urban transformation needs to be qualified. The new processes of urbanisation that have been recognised as the causes of this transformation actually produce a geographically uneven pattern of 'transformation' which varies regionally, between cities and within cities.

Further reading

Several recently published books have looked at 'new' processes of urbanisation through a number of international case studies. While all adopt different slants some of the best have included:

Brotchie, J., Batty, M., Blakely, E., Hall, P. and Newton, P. (eds) (1995) *Cities in Competition: Productive and Sustainable Cities for the 21st Century* Melbourne: Longman Australia.

Knox, P.L. (ed.) (1993) *The Restless Urban Landscape* Englewood Cliffs, NJ: Prentice-Hall.

Watson, S. and Gibson, K. (eds) (1995) *Postmodern Cities and Spaces* Oxford: Blackwell.

An introduction to edge-city development in Europe can be found in the special edition of *Environment and Planning D: Society and Space* (1994) vol. 12 no. 2.

An interesting discussion of the impacts of electronic technologies on urban form can be found in:

Graham, S. and Marvin, S. (1996) *Telecommunications and the City: Electronic Spaces, Urban Places* London: Routledge (Chapter 8).

7 Transforming the image of the city

- The significance of urban image
- Place promotion and the post-industrial economy
- Place promotion and urbanisation
- New urban images
- A critique of place promotion

The previous chapters explored some of the strategies that have commonly been employed since the early 1980s in the UK, mainland Europe and North America to regenerate the landscapes and economies of towns and cities. In this chapter the focus shifts to consider the transformation of the image of the city and the negative images that have become associated with many cities following the deindustrialisation of their economies and which have proved a hurdle to their successful regeneration.

What is an urban image?

All cities have an image. In fact, it would be truer to say that all cities have, and always have had, a number of images. A place image of any kind is the simplified, generalised, often stereotypical, impression that people have of any place or area, in this case of cities. Yet it is impossible to know cities in their entirety. To make sense of our surroundings we reduce the complexity of reality to a few selective impressions. In being selective in this way we are producing a place image. Place images typically exaggerate certain features, be they physical, social, cultural, economic, political or some combination of these, while reducing or even excluding others. That the actual conditions in a city may have changed considerably since the image of that place was formed is not the point. In the world of perception the image is more important than the reality. This last point might prove to be both an advantage and a disadvantage to cities. On the one hand, it means that an urban image can be cleverly manipulated and transformed by city marketers without the trouble of having to affect actual substantial change in that locality; on the other hand, it means that negative and increasingly misleading images may persist despite considerable change having taken place.

Forming urban images

Since the mid-1980s a large, and apparently growing, industry has developed around the deliberate manipulation and promotion of place images, which has become an integral part of urban regeneration programmes. This is explored later in this chapter. However, promotional campaigns by local authorities are not the only methods by which images of places are formed. Persuasive urban images can be formed in a variety of other ways. These can involve:

- media coverage of events in places which become the prevailing impressions of those places (the riots in various British inner-cities during the 1980s and in Los Angeles in 1992 would be examples of this)
- satire (comedians such as Billy Connolly and television comedies such as *Rab C. Nesbit* have long used Glasgow's reputation for drunkenness and violence as a staple of their comedy)
- personal experience (visits to cities as, for example, tourists are frequently of short duration and by necessity highly selective, focusing on sites of interest or appeal and excluding large areas of cities)
- hearsay and reputation (what people tell us about cities, whether from personal experience or hearsay, forms an important component of our impressions of places).

Bad urban images

Many cities have suffered the stigma of a bad image both now and in the past. One might think of the examples of Liverpool's reputation for crime, Manchester for drug problems, Los Angeles for social and ethnic conflict and Birmingham for being an architectural and cultural wasteland. Clearly these are only partial truths in each case as Liverpool does not have a monopoly on crime nor Birmingham on poor architecture. This is not the point. The images persist. Bad urban images tend to derive from the exaggeration of elements from a poor physical environment, a narrow and restricted cultural profile, social polarisation and unrest, economic dereliction and depression.

At this point it is important to say that urban images are not inherently good or bad in themselves. It is their relation to wider cultural fashions and trends that determines the ways in which they will be regarded. For example, during much of the twentieth century industry was a positive image for a city to possess. Changing fashions, linked to changes in the global economy discussed earlier, have caused industry to become regarded in a negative light. While once industry was equated with power, skill and pride, it is now more likely to be associated with dereliction, economic decline and pollution (Short *et al.* 1993). This change is reflected in the ways in which it has tended to be distanced from cities in their more recent promotional campaigns.

The rise of urban place promotion

The promotion of urban places has long been an integral aspect of urban development and hence urban geography. This is something that conventional accounts of the discipline has only recently recognised. Place promotion has affected the development of a wide range of urban environments in Britain, North America, Europe and Asia. It has involved a diversity of private and public institutions, the latter including all levels of local, regional and central government, and it has displayed distinctly different histories in different parts of the world (Ward 1994).

Some of the earliest examples of place promotion were found in the underpopulated west of North America from the early nineteenth century. Place promotion was used here to try and sell real estate at a time when many towns were being established (Holcomb 1990; Ward 1994). Place promotion in a form recognisable in the 1990s emerged in the north-eastern USA in the mid-nineteenth century and involved the promotion of industrial towns. Later in the century this promotion spread to the industrial provinces of Canada. Initially this type of promotion was largely a local activity which included institutions such as municipal boards of trade, councils and chambers of commerce, often in conjunction with private developers, railway companies and local fiscal initiatives. The main promotional publications to emerge from this were local newspapers, business and trade directories and promotional brochures. However, over the course of the twentieth century in Canada and the USA, place promotion began to involve progressively higher levels of state, provincial and federal government, as it became integrated into regional development programmes (Ward 1994: 54–62).

Place promotion in Britain and Australia displayed very different histories from those of the USA and Canada, and focused, initially at least, on the promotion of different types of urban environments. Place promotion first emerged in Britain in the mid-nineteenth century with the promotion of mass tourist coastal resorts such as Blackpool and Scarborough (Ward 1988) initially by railway companies eager to drum up extra business, but later by the towns themselves. Promotional activity in the early twentieth century shifted to the promotion of the expanding residential suburbs, especially around London. Again railway companies were major agents of this promotion, but now they were joined by private developers and building societies (Gold and Gold 1990; 1994). It was not until the 1930s that the promotion of industrial towns occurred on any great scale in Britain (Ward 1990; 1994). This was some fifty years after it had become widespread in the USA.

In Australia promotion was largely a post-war activity which formed a cornerstone of Australia's policy to attract immigrants and address its underpopulation problems. The main agent in this effort was national government, although local initiatives from individual towns and cities were also important (Ryan 1990; Teather 1991). Australia has also been marketed as a tourist destination by airline, travel and holiday companies and Australian tourist organisations.

Despite these early examples, place promotion has assumed far greater importance since the early 1970s, when waves of deindustrialisation generally affected the UK, Europe, North America and particularly affected former industrial cities. First, more cities are recognising the need to promote positive images of themselves than was previously the case. It is now the exception rather than the rule to find an urban area not engaged in vigorous promotional activity of some kind (Table 7.1).

Second, the images produced by cities are more diverse than in previous years. Rather than simply promoting a unitary image of the city, industrial cities are recognising that they are not catering for a single,

Table 7.1 *Local authority promotional packages 1977 and 1992*

	% of local authorities	
	1977	1992
Guide	42.6	84.2
Glossy	29.7	56.2
Fact sheet	20.3	37.7
Industrial/commercial information	20.9	69.9
Tourist	28.4	84.9
Other	42.6	85.6
Slogan	43.9	45.2
Magazine/newspaper	—	32.2
Coat of arms	—	36.3
Logo	—	73.6

Source: Barke and Harrop (1994: 97)

homogeneous audience, but for a plethora of distinctive niche markets. Consequently they tend to put forward a variety of images of themselves. Another aspect of this diversity is that a far wider range of organisations are involved in the business of place promotion. While previously a chamber of commerce, city corporation or local authority might be responsible for place promotion as well as a few private businesses such as railway and travel companies, the groups involved in the 1990s are more diverse. Some of these groups might be partly or wholly co-ordinated by a specialist national body (such as the Great British Cities Marketing Group) or local organisation (such as Birmingham's Marketing Partnership); however, they might also include central government agencies such as Urban Development Corporations, departments of the local authority, specialist facilities such as convention centres and airports, local institutions such as universities, which may have initiated schemes like science or research parks, members of the local business authority and the local media. The effect produced is a collage of often contrasting, but appealing, images of cities.

Finally, spending on place promotion, particularly as a proportion of local authority budgets, has increased since the early 1980s. This expenditure also forms a significant proportion of spending by Urban Development Corporations and other agencies concerned with regeneration.

Despite these diverse histories the widespread and international process of deindustrialisation has seen the development of much homogeneity in the types of promotions undertaken by cities and consequently the types of images produced. Although this is not to say alternatives do not exist, it is an inevitable reflection of the increasingly interconnected global economy.

Urban image in the post-industrial economy

A number of changes have occurred in the organisation of the global economy that have made the promotion of a positive image of place an extremely important part of economic regeneration. These were discussed in detail in Chapters 3 and 4. To summarise, it can be seen that the networks of competition that individual cities find themselves caught within have increased both spatially and numerically, and individual cities are subject to fewer protective measures and structures than has previously been the case.

In the UK, Europe and North America, many of the formerly most prosperous cities have suffered waves of deindustrialisation in their economies. They have broadly recognised that they might tap the growth displayed by the tertiary or service sector of the economy. However, these sectors are very different in their character from the secondary, manufacturing sector from which these cities' wealth was, in the large part, derived. The locational requirements of this sector included close proximity to raw materials, a large supply of labour, good transport links and accessible markets. Also a large amount of capital tended to be tied up in industrial plant. For profits to be generated this plant was required to generate economies of scale. These were achieved through large-scale production, usually of a uniform product for a long period. Consequently, this sector displayed a high degree of geographical *inertia*. They were locationally tied to specific types of sites and, once established, needed to remain there for long periods to generate production economies of scale.

By contrast the growth areas of the economy now tend to display much less geographical inertia. The traditional locational requirements are now no longer as important than was previously the case. Cities, therefore, need to establish new advantages for themselves. Service sector activity involves little or no capital being tied up in heavy machinery or factories. The boom in the property development office rental sectors in the 1980s mean that there was a ready supply of appropriate space in almost every city. The growth in the development and employment of electronic communication technology has meant that markets can be instantly accessed from anywhere in the world. The traditional geographical attributes of location appear to be becoming an increasingly irrelevant aspect of the locational decision-making of service sector activity. Firms in the service sector are considerably more footloose than those in the secondary sector.

This has had important implications for the economic regeneration of cities. They have recognised that firms and investment now respond to a very different set of circumstances than previously. It appears that investment decisions in these growing sectors, given their different locational requirements, respond more to differences in image than to other more tangible locational factors.

> How an area is perceived and its physical or environmental desirability, however notional, will affect the levels of investment by industrial property developers, financial interests, and companies, on the one hand, and the inclination of employers and employees to work and live there, on the other.
>
> (Watson 1991: 63)

Footloose inward investment provides a potentially very unstable basis for urbanisation because it is able to switch location to a much greater degree than manufacturing or heavy industrial investment, to respond to even very slight changes in conditions elsewhere. This instability is exacerbated by three related conditions in the nature of urban economies and international economic systems. First, inter-urban networks of competition have become increasingly international in nature. Because of the global nature of many economic systems and markets, for example the business tourist and convention market, the choices of location available to decision-makers are potentially world-wide. Therefore, cities now commonly find themselves in competition with cities, not only from their own country, but, frequently, from all over the world.

Second, not only has the spatial extent of networks of urban competition increased but also so has their numerical extent. Cities have traditionally displayed some degree of specialisation in their economic profile. While this certainly did not immunise them from competition it at least tended to reduce the number of cities with whom they were in competition. Networks of competition tended to develop among cities with similar economic profiles. However, cities found that these layers of economic specialisation were stripped away by the waves of deindustrialisation that affected them from the early 1970s onwards. The strategies adopted by cities to regenerate their landscapes and economies have tended to reduce this specialisation further. Urban regeneration projects in cities of the UK, Europe and North America have displayed a remarkable degree of similarity. The consequence of this is that cities have tended to find themselves in competition with an ever increasing number of cities entering the growth sectors of industrial and office relocation, business tourism and cultural tourism, as well as established cities that have traditionally dominated these sectors. This has raised problems not just at the local urban scale but at the national scale too. These have included the problems of market saturation and zero-sum growth at a regional or national scale, namely, the problems associated with growth in one area resulting in decline in another.

Third, in many cases in the past, the economic position of many cities was secured, or at least protected, by protective structures or agreements. These included government regional policy, trade agreements between countries, military force or occupation and systems of empire. While these have far from vanished entirely from the world economy they have certainly been reduced in number and scope and have changed in nature. Central government regional policy, for example, has been replaced by grants such as the European Commission's regional fund and City Challenge in the UK, which are decided by a process of competitive bidding between areas. These types of awards have tended to increase the climate and extent of inter-urban competition and the insecurity of these types of regional investment.

The strategies of economic and urban regeneration employed by many formerly industrial cities in the UK, mainland Europe and North America appear to be caught within an increasing spiral of insecurity and competition. On the one hand, the markets they have and continue to enter have been characterised by an apparently increasing degree of competition. On the other hand, historically, whenever urban systems have displayed increased instability, the response of individual cities has been to increase their speculative urban regeneration programmes and associated promotional activities. This vicious circle of instability appears difficult to break. Certainly, the examples where cities have sought innovative responses to attempt to break it are rare. The majority simply fall back onto the, now rather unoriginal, routes of spectacular property development and urban promotion. While individual schemes may be judged as innovative they are part of a wider strategy which forms an increasingly problematic basis for urbanisation.

Urbanisation and place promotion

The processes of place promotion are not incidental to those of urbanisation. As image assumes ever greater importance in the post-industrial economy it is becoming clearer that the actual production of urban landscapes reflects the necessity for cities to present positive images of themselves and that economic development is driven by programmes of place promotion. It is important to appreciate the distinction between 'selling' and 'marketing', for the two are very different processes. Selling is a process whereby consumers are persuaded that they want what one has to sell. However, marketing is a process whereby what one has to sell is shaped by some idea of what one thinks the consumer wants (Fretter 1993: 165; Holcomb 1993; 1994). The distinction between 'selling the city' and 'marketing the city' is, therefore, crucial to understanding their relationship with urban development. 'Selling' the city is likely to impact most directly on the urban economy through increased visitor spending and outside investment. However, marketing cities also impacts directly upon their landscapes and development. It would be true to say that prior to the 1970s cities were largely 'sold'. It is now truer to say that they are 'marketed'. City landscapes are increasingly shaped according to views of what potential consumers want. Marketing the city, therefore, is a process that is increasingly integral to the shaping of urban development, rather than incidental to it.

> [Place marketing] is the principal driving force in urban economic development in the 1980s and will continue to be so in the next decade. . . . The logic that more jobs make a better city is giving way to the realisation that making a better city attracts more jobs.
>
> (Bailey 1989: 3)

> 'Marketing' is starting to replace the concept of merely 'selling'. . . . Selling is trying to get the customer to buy what you have, whereas marketing meets the needs of the customer profitably (or, in local government terms 'efficiently' at best value for money). Inevitably, this requires a much more sophisticated and more comprehensive approach affecting many local authority functions. Place marketing has thus become much more than merely selling the area to attract mobile companies or tourists. It can now be viewed as a fundamental part of planning, a fundamental part of guiding the development of places in a desired fashion.
>
> (Fretter 1993: 165)

The process of marketing cities

The ways in which the city has been promoted have been diverse. These have included distribution of guides, brochures and other information through tourist offices, libraries and commercial information services, through responses to postal enquiries, poster advertising, particularly in sites where large numbers of their intended audience are likely to gather, for example, major railway stations and airports, through press advertisements, especially in the financial and property pages of the broadsheet newspapers, in specialist property or commercial pull-outs, or in specialist magazines, and through the employment of recognisable slogans and city logos. Increasingly cities are taking a proactive view of their marketing. Cities are now frequently sending out representatives, from the local authority and business community, to meet potential customers overseas. Throughout the late 1980s and early 1990s, prior to the opening of its International Convention Centre, Birmingham City Council ran a 'Birmingham roadshow' which toured Europe, and more distant markets like Japan, meeting potential customers and advertising the city through a series of face-to-face meetings.

The market for place promotion

Cities do not simply market themselves indiscriminately, but carefully target specific audiences, which include companies in the expanding service sectors of the economy and organisations involved in the planning of business tourist events like conventions.

These audiences have an important determinate effect on the images that cities promote. The service sector does not tend to require a large unskilled labour force but a smaller, highly qualified and highly skilled labour force of, largely, middle-class professionals. Likewise business tourism usually involves middle- or high-ranking representatives of companies. The values that this audience is seen to respond to are highly specific and involve a concentration on lifestyle issues, including culture and environment, as well as business issues. While the markets may be broadly similar, cities are attuned to their subtle differences and direct their marketing accordingly.

New images for cities

A cursory survey of promotional literature from almost any town or city will reveal that they are keen to promote themselves as a good place to live as well as a good place to work. Cities emphasise not only their business opportunities but also their lifestyle activities (see Case Study I).

Case Study I

Marketing North American post-industrial cities

North American cities, despite their diversity, suffered from a number of common problems during the 1980s. The severity and extent of these problems, while varying between individual cities, were generally felt most severely in the rust-belt of predominantly north-eastern, formerly industrial cities. Their problems evolved from a coincidence of deindustrialisation with a number of other detrimental social, demographic, political and fiscal changes, including the suburbanisation of wealthy populations, frequently taking them beyond metropolitan boundaries, at the same time as minority and low income populations were occurring near the city centre. This, combined with a cut in support from central government, plunged many cities into fiscal crises. The responses of North American cities to these problems were an increasingly entrepreneurial stance by local authorities who actively sought to promote and initiate development through public–private partnerships, spectacular redevelopment of their downtowns, vigorous programmes of place promotion through advertising, redesigning logos and chasing sporting events and franchises. However, during a time of heightened inter-city competition when it has become imperative for cities to promote some claim to distinctiveness, their efforts have led to an increasing homogeneity of city landscapes and images.

This can be accounted for in part by the underdevelopment of place marketing in North America and its institutional structure. Despite the vast increase in the emphasis placed on marketing by North American cities during the 1980s, compared to the marketing of many commercial products, it appears that cities remain under-marketed. Estimates of the expenditure by cities and states on place marketing vary. One estimate put 1986 expenditure in the $3 billion – 6 billion range (Guskind 1987: 4) although the late 1990s figure is likely to be higher. However, in comparisons to the marketing of many commercial products this figure is very low. The institutional structure of place marketing is similarly poorly developed. There are very few private companies who specialise in place promotion, the majority of work is conducted in-house by departments of government and local authorities in association with advertising firms who are commissioned for specific projects. Consequently, much work in this sphere is undertaken by people with no professional background in marketing. The campaigns thus produced are frequently rarely original or well designed (Haider 1992). These campaigns tend to lapse into the reproduction of a few standard packages. Similarly the effectiveness of campaigns is rarely scrutinised.

Analysis of the promotional packages produced by North American cities reveals the limited range of the themes and motifs which they employ.

Theme	Example
The future	'New Brunswick Tomorrow' 'Growing Strong' [Connecticut] 'A Great Past and a Greater Tomorrow' [New Milford, Connecticut]
Centre	'Put your Business in the Centre of the World' [Atlanta] 'We're in the Middle of Everything' [Corona, California] 'At the Centre of the World' [Hempstead, New York State]
Gateway	'Your Gateway to Mexico' [Brownsville, Texas]
Cheap Location Costs	'Consider the Less Taxing Environment' [Orlando, Florida]
Labour Quality	'Empora: Working for You' 'Twin Fall Idaho Works'
High Quality of Life	'A Liveable Place to Work' [Santa Maria, California] 'The City that Exceeds Expectations' [Chicago] '#1 Most Liveable City' [Pittsburgh]

Source: Holcomb (1994)

In the successful promotion of place the two are seen as indispensable and intricately intertwined. This section will explore in detail the images that cities have sought to create for themselves and promote as part of their economic regeneration.

Centrality

Now, as always, cities are desperate to create the impression that they lie at the centre of something or other. This idea of centrality may be locational, that a city lies at the geographical centre of England, the UK, Europe and so on. This draws on a deep-seated notion that geographical centrality makes a place more accessible, easing communication and communication costs. However, now that the economy is characterised more by the exchange of information than hard goods, geographical centrality has been superseded by attempts to create a sense of cultural centrality.

From Exhibitions to Entertainments to Conferences to Conventions to Sports to Spectaculars...

We're at the Centre of Everything.

It's easy to see why Birmingham is becoming known as the Exhibition and Convention City.

Where else could offer such diversity – from huge trade fairs and exhibitions to international conferences, world class music and top sporting events?

While each of the four venues has been built for a specific purpose, they can also be used for almost any event you wish to stage.

But whichever you choose you can be certain it'll be at the Centre of Everything.

Plate 7.1   *Magazine advertisement for Birmingham*
Reproduced with the kind permission of NEC Group Ltd

Cultural centrality usually manifests itself as a cry that a city is at the centre of *the action*, meaning that the city has an abundance of cultural activities, such as bars, restaurants, night-clubs, theatre, ballet, music, sport and scenery (Plate 7.1). The suggestion is that people will want for nothing in this city.

Images of industry

Since the mid-1980s former industrial cities have been among the most vigorous place promoters. Their industrial history and the process of deindustrialisation have created something of an image problem for them. First, the most widely held images of industry are negative ones. They are of heavy, dirty, dangerous, polluting, rough, working-class activities. The process of deindustrialisation has created the images of dereliction, economic decay and unemployment – hardly the things to attract the well-heeled executive to a city. Promotional activity has concentrated on replacing these images of formerly industrial cities and changing the image of industrial activity generally.

> To call a city 'industrial' in the present period in the U.S. is to associate it with a set of negative images: declining economic base, pollution, a city on the downward slide. Cities with more positive imagery are associated with the post-industrial era, the future, the new, the clean, the high-tech, the economically upbeat and the socially progressive. We can identify a number of polarities in the division between industrial and post-industrial. Industrial cities are associated with the past and the old, work, pollution and the world of production. The post-industrial city, in contrast, is associated with the new, the future, the unpolluted, consumption and exchange, the worlds of leisure as opposed to work.
>
> (Short *et al.* 1993: 208)

The landscapes within which images of industry are framed are important in attempting to change the traditional images of industry. These landscapes are not those of the factory, the smokestack chimney and the storage yard, but are the highly designed, green landscape of the science park. These typically contain well-maintained, landscaped lawns and gardens as well as lakes and sculptures. The buildings that populate this landscape are the post-modern, ornate, futuristic buildings of light, hi-tech, clean industry (Plate 7.2).

The promotion of actual industrial activity is similarly sanitised. The images of the production process contained in promotional activity

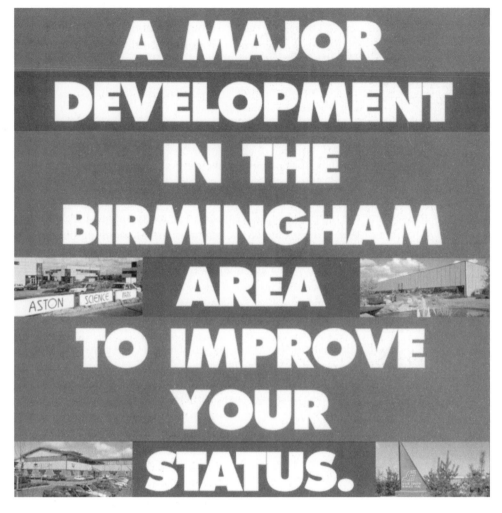

Plate 7.2 *Promotional images of industry*
Source: Birmingham City Council

specifically seeks to change the image of industrial activity as heavy, dirty and dangerous. The overriding impressions created by these images are of technology, skill, precision, and cleanliness.

Together these images aim to replace the traditional negative images of industry with more positive, appealing ones. These images suggest that industry represents the successful marriage of human skill, technology and environment, a far more appealing set of associations.

Plate 7.3 *The Black Country Living Museum, near Dudley, West Midlands*
Source: The Black Country Living Museum

Industrial heritage sites have formed an important part of many city's tourist strategies. This is so for a number of reasons, including the presence of large areas of prime central city property left empty following the decline of, for example, canal transport or dock activity. As well as restoration creating a series of desirable landscapes for the professional, industrial heritage can be used to generate positive images of places. This industrial tourism has been criticised because of the heavily idealised, sanitised treatment that it gives the past. Certainly, restored mills, docks, canals and heritage centres depict workers as cleaner, healthier and happier than they were ever likely to have been in reality (Plate 7.3).

The negative social attributes of the industrial past – the male dominance of employment and public life, or the role of ports in the slave trade, for example – are largely absent from modern representations (Case Study J). These images of industrial heritage play upon positive images such as pride, innovation, skill, strength and tradition.

Images of business

Dynamic images of places as good business locations have become a cornerstone of modern urban promotion. Rather than being organised around mundane functional attributes the promotion of locations for

Case Study J

Syracuse, New York: a tale of two logos

Syracuse is a city of 164 000 (1990) located approximately half-way between New York City and Toronto. Its economy, which grew for almost a hundred years until the 1970s, was based initially on chemicals. From the early twentieth century it diversified, becoming a major world manufacturer of bicycles and typewriters. It was known as an 'industrial' city. However, since the 1960s the city has been affected by a slump in manufacturing which became particularly severe during the mid-1980s. Syracuse's original logo (Plate 7.4a), which dated from 1848, reflected both the dominance of industry within its own economy and the positive image associated with industry more generally. The old logo shows saltfields and smokestack chimneys. However, by the 1980s, industry had become associated with a whole range of

Plate 7.4b *Syracuse logo 1986*

negative images. In 1986 a competition was held to design a new civic logo. The winner not only reflected the shift in attitudes towards other aspects of the economy and the environment within Syracuse, but also, being designed to project a positive image to an external audience, reflected a wider shift in attitudes. Reflecting the rise of environmentalism within North American society, the new logo showed the newly valued aspects of Syracuse's environment, the recently cleaned-up lake and a modern, post-industrial skyline (Plate 7.4b). Excluding the industrial legacy of the city in representations such as this illustrates the shift of attitudes away from the industrial. Redesigning its logo in this way was one way in which Syracuse effected the transformation of its image from industrial to post-industrial.

Plate 7.4a *Syracuse logo 1848*

Source: Short *et al.* (1993)

business have been dramatic and highly visual. They have focused primarily on images of architecture, communication and technology.

The spaces within which business is conducted, such as office towers, convention centres and business parks, have become crucial to the projection of city image and status. Buildings are important flagships of urban regeneration (see Chapter 6). Their size, design and capability are major icons of prestige. Consequently, the images of business play heavily upon the architectural setting and in doing so perpetuate the role of the building and architecture in defining city status (Plate 7.5).

Increasingly business is being conducted on a global scale. Electronic communications have allowed the instant exchange of information between vastly separated regions. This ability to conduct business globally has become another very important signifier of city prestige and status. Promotional images attempt to create the impression that their cities have the capability to annihilate the restrictions of space from a regional to a global scale, through physical transport (road, rail and air) and electronic communication. The prevailing image of cities as locations for business are of powerful nodes shaping a *global* economy, regardless of their actual position. These images are vital if cities are to attract investment and relocation from progressive businesses with international aspirations.

The suggestion of the presence of specialist and quality skills within localities is important, given that the labour requirements of business are ever more qualitative than quantitative. Consequently promotional materials emphasise the links between the industrial and business communities and the educational communities. These images act as signifiers of quality and excellence.

Images of lifestyle

Urban promotion involves the selling of a location not only for business, but also as a place to live. It is vital, if people 'of the right sort' are to be attracted to an area, to suggest that it is equally able to offer lifestyle as well as business opportunities (see Case Study K). These images of lifestyle tend to be predominantly anchored around two things: culture and environment. The use of leisure time is considered an increasingly important aspect of the decision-making process for both long-term relocation decisions and short-term (for example, convention location decisions) business or tourist decisions.

Plate 7.5 *Promotional images of business*
Source: Birmingham City Council

Case Study K

Changing the image of London's Docklands

The physical regeneration of London's Docklands was underpinned by a series of concerted efforts by both the London Docklands Development Corporation (LDDC) and private developers to promote a positive image for the area. The response has been to create a number of complementary images for the area. London's Docklands has been promoted as a number of things: a place to invest in; a place for businesses to relocate to; a place for the affluent to work, live and play and a major tourist destination. The efforts to transform the image of the area have been wide-ranging and have encompassed media advertising campaigns, the careful development of the built environment and a number of arts-related initiatives. Prior to the 1980s Docklands was perceived as an area blighted by industrial decay and dereliction. To counter these images the area was primarily promoted as a place with a future. This was most evident in the variety of slogans attached to it, including 'watercity of the twenty-first century' and a 'triumphant urban renaissance'.

The LDDC was responsible for two advertising campaigns. The first, during 1982–3, was a television and radio campaign aimed at raising the awareness of the existence of Docklands and the developments going on there. It played upon images of the area's centrality as a site for business relocation. The second campaign was a newspaper campaign in 1990 which advertised more specific attractions of Docklands. Both campaigns stressed the uniqueness of the area. The images of Docklands promoted in these campaigns were given some substance by the careful shaping of its built environment. Two motifs dominated this process: heritage conservation and spectacular architecture. The retention and renovation of Docklands' historical architectural legacy was central to the physical regeneration of the area. Docklands contained seventeen conservation areas and hundreds of listed buildings and structures.

Source: Crilley (1992)

The notion of culture that cities have used within their promotional strategies has been very narrow and consists entirely of what might be termed high culture. This typically consists of things like theatre, ballet, classical music, art galleries and museums. Where more popular activities are included they still reflect activities that are likely to appeal to a fairly narrowly defined target audience of affluent middle-class professionals. These include exclusive wine-bars, restaurants and pubs, designer shopping facilities, cinemas and night-clubs. They suggest an abundance of exclusive or refined leisure facilities.

Images of environment

The essential complement to this cultural image is an equally appealing environmental image. Traditionally the environment of cities has been perceived very negatively. Since the mass industrialisation and urbanisation of the nineteenth century the image of the city as dirty, overcrowded, unfriendly and unhealthy has tended to dominate images of the city, in the UK and North America at least. Added to this, more recent images of economic recession, environmental decay and social unrest in inner cities and on peripheral housing estates have not helped to enhance the appeal of the city. Clearly cities still suffer from an image problem, some more so than others.

To counter this, local authorities have sought to re-invent or 're-imagine' the urban environment in their promotional campaigns. This has been achieved through the construction of a collage of positive environmental images, which typically consists of three elements – architecture, suburbs and countryside.

Architecture

Images of this tend to be of two kinds, spectacular and historic. Spectacular, futuristic, ornate, post-modern architecture such as flagship projects of urban regeneration suggests a progressive, dynamic city on the move. Images of historic architecture suggest civic tradition of some kind such as local government (town halls) and art (galleries, concert halls).

Suburbs

These images are very domestic. They consist of prestigious, detached houses, manicured lawns, well-kept gardens, and suggest a 'nice', safe place to live.

Countryside

The presence of picturesque countryside near at hand is frequently emphasised in promotional materials. It offers an escape from the city and somewhere to relax and for recreation (golf, sailing, climbing and walking).

How successful is urban promotion?

It is difficult to generalise about the success or the influence of place promotion campaigns in affecting the relocation and investment decisions. The reactions of managers and decision-makers range from those who argue that advertising is incidental to their decisions, which are based on sound economic grounds, to those who argue that advertising is crucial to shaping the perceptions of their workforce and their clients. It is likely that in certain sectors such as conference, convention and exhibition organisation, the images projected by location do play an important part in the decision-making process. Despite doubts over its effectiveness, the sheer amount of place promotion occurring is likely to ensure its continuation. As competition between urban places for a whole range of public and private investment continues to intensify, it is likely that if places fail to advertise their claims, they will simply not get noticed.

Conclusion

> [City] councils which seek to promote a dream image of their city not only ignore the social consequences of focusing on private sector needs and perceptions: they can actually make matters worse for poor people.
>
> (Hambleton 1991: 6)

The images of the city created by promotional campaigns are highly selective. These images have been based around the emphasis of certain positive images and the exclusion of unappealing, negative images. It has been argued that this process is not innocent. It has been shown that it is an important part of the contemporary urbanisation process. It has been argued that in marketing cities towards a small, wealthy elite, the needs of less well-off people have been ignored. Some geographers have argued that the positive images of the city created in promotional campaigns and in urban regeneration have acted as 'masks' hiding the reality of urban problems (Harvey 1988: 35).

As with any process of urban change the benefits and problems are not evenly distributed. The preceding chapters have considered how cities have redeveloped and regenerated their economies, landscapes and images to cope with the stresses of deindustrialisation. Chapter 8 considers the new social and cultural geographies etched out by this process.

Further reading

A diverse range of up-to-date essays on many aspects of urban promotion can be found in:

Gold, J.R. and Ward, S.V. (eds) (1994) *Place Promotion: The Use of Publicity and Marketing to Sell Towns and Regions* Chichester: Wiley.

Kearns, G. and Philo, C. (eds) (1993) *Selling Places: The City as Cultural Capital, Past and Present* Oxford: Pergamon.

The promotion of cities for tourism is explored in:

Ashworth, G.J. and Voogd, H. (1990) *Marketing the City: Marketing Approaches in Public Sector Urban Planning* London: Belhaven.

Page, S. (1995) *Urban Tourism* London: Routledge (Chapter 6).

8 New geographies of old cities

- The myth of urban renaissance
- The impacts of urban regeneration
- Future urban regeneration

It should be apparent that cities have been forced to adjust to severe structural changes occurring within national and international economies. Some of the most heavily affected are the former manufacturing cities of various rust-belts, in the north of the UK, north-east USA and certain Australian states (see Chapter 3). These cities have been at the sharp end of the effects of the globalisation of the world economy and its transition from industrial to post-industrial. They have seen their economic bases largely eroded and with them their former affluence and economic pre-eminence. How they have adjusted to these wider economic changes is of profound importance to their current and future welfare and development. This chapter offers an evaluation of the measures undertaken by these cities to adjust to changing economic circumstances. Chapter 9 sketches out some alternative scenarios of how this adjustment might be negotiated into the twenty-first century.

Evaluating urban regeneration

Previous chapters have outlined various ways in which the management of cities has broadly become more entrepreneurial and enterprise oriented. The focus of urban government has shifted away from managerial, welfare issues towards those of promoting economic development through regeneration. This tendency has been particularly well developed in many British, European and North American rust-belt cities. 'Regeneration' should be regarded as a problematic term and consequently viewed critically. The question 'what does regeneration mean?' is an important one yet is ill considered by many politicians and policy-makers. All too often regeneration is equated with economic or physical regeneration. This in turn is equated with an increase in wealth or jobs in an area. However, this is an inadequate, or at the very least, narrow definition of regeneration. This definition ignores a number of issues which are crucial to the evaluation of the success or failure of projects of urban regeneration.

The first issue concerns the characteristics of this wealth or these jobs. For example, what is the distribution of wealth among different social groups? Is it equal or fair? What are the type and quality of jobs stemming from urban regeneration? What is the distribution of these jobs between different social groups? The second issue concerns whether or not all of the impacts of urban regeneration are positive (see Case Study L). Are there any negative impacts of urban regeneration? If so, how are these negative impacts distributed between different social groups?

The aim of these questions is to open up the claims of urban regeneration and its rhetoric to critical analysis. This chapter evaluates the claims of urban regeneration and assesses its economic, social, political and cultural impacts on cities and their populations.

Economic impact

At first glance spectacular projects of urban regeneration appear to offer attractive alternatives to the derelict industrial land that many are built upon. As conspicuous displays of wealth and with their emphasis on high-quality urban design, they create the impression of revival. They are confident architectural statements. They also offer a number of direct and indirect benefits to the urban economy. By being 'people-attractors' of some description these projects lure a significant number of visitors to cities. Bradford's National Museum of Film and Television is a good example. This development has been instrumental in virtually creating a tourist economy in Bradford out of nothing (Bianchini *et al.* 1992: 251). Recorded visits to Bradford increased from none in 1980 to well over 5 million ten years later. The beneficial spin-offs from this cannot be underestimated. Spending in the city, and others such as Glasgow which have adopted similar approaches to regeneration, has increased significantly in retailing, entertainment,

Case study L

Failure of regeneration in Glasgow

Glasgow's economy was crippled by deindustrialisation in the 1970s and the early 1980s. Its response to this crisis was to enthusiastically embrace a series of culture-led urban regeneration projects. These projects included the 'Glasgow's miles better' promotional campaign, the 1988 Garden Festival and the year long 'European City of Culture' celebrations during 1990. In the early 1990s Glasgow was held up as the model of how former manufacturing cities can successfully transform their landscapes, images and economies through cultural urban regeneration. Glasgow had the appearance of an affluent, confident, 'cultural' city.

However, four years after its apparent zenith as the European City of Culture, critics were beginning to question the substance of Glasgow's apparent transformation. The atmosphere in the city, by the mid-1990s, was somewhat less confident. The reason for the city's failure of confidence was the persistence, in some cases worsening, of some severe social and economic problems in Glasgow. The figures are startling. In 1994 one-third of the city's population still depended on income support, more than three times the UK's national average; one in four of Glasgow's male population was unemployed; women in Glasgow suffered the worst diet in the Western world; Glasgow's peripheral housing estates had the highest proportion of injecting drug addicts per head of population anywhere in Western Europe.

The persistence of these problems provided a persuasive argument that, if Glasgow's regeneration was not to be regarded as a failure, then its successes have been extremely limited. A large proportion of its population, one-third or more, have failed to reap the benefits of the city's apparent transformation. While Glasgow certainly created the appearance of regeneration the substance of this regeneration was less certain.

Source: Arlidge (1994)

catering and hospitality, hotels and other consumer service industries. As a result the benefits of regeneration have spread throughout the city centre, transforming the appearance of many older and formerly derelict buildings. Old buildings have been renovated and refurbished and new ones built to tap into the expansion of the tourist economy. Moreover, one of the prime functions of these types of development has been to raise the national and international profile of towns such as Bradford or at least ensure that they are known for positive rather than negative reasons (Bianchini *et al.* 1992: 249–51). City leaders and promoters refer to this as 'putting the city on the map'. This has been important in attracting the attention of national and international organisations such as national tourist boards and the European Union, particularly where grants or financial help has been available to subsidise these developments. However, benefits such as these should be balanced against the complex, multifaceted and interrelated nature of problems in the city. The example of Glasgow demonstrates the limits of regeneration programmes. An evaluation of the impacts of urban regeneration should ask: how and why does regeneration fail? And who does regeneration fail?

To recap, the economic rationale behind regeneration is for a redistribution of income within the city via a combination of 'trickle-down' and the multiplier effect. The assumption, for there is little actual evidence of its effectiveness, is that increased revenue from visitor spending and investment will do two things. First, it will trickle-down into the pockets of the most disadvantaged through the creation of jobs servicing visitors and incoming investors (Hambleton 1995a). Second, it is assumed that this revenue will have a positive, knock-on effect as it spreads through the local economy through increased consumer spending. These assumed economic mechanisms are held up as justification of the claims of urban regeneration.

However, this rhetoric, and the inherent assumptions, ignore a number of critical issues. Perhaps most importantly they ignore the compelling evidence against the effectiveness of the trickle-down and multiplier mechanisms. Further, they ignore the wide-ranging negative impacts that projects of urban regeneration can have on the local economy and upon certain groups within it. Clearly in any genuine evaluation of the impacts and effectiveness of urban regeneration, rational debate needs to replace the rhetoric that has been characteristic.

As mechanisms of income redistribution advantaging the poorer sectors of urban societies, both the trickle-down and multiplier effects are likely to be very ineffective. The evidence suggests that either the increased revenue in urban economies as a result of urban regeneration will fail to 'drain' out of the business and managerial sectors or it will do so through those channels that fail to benefit poorer populations substantially. These channels include the creation of low-paid, unskilled, insecure, part-time employment. These are likely to be the only employment opportunities stemming from projects of urban regeneration that are open to poorer populations (Loftman and Nevin 1994). The economic effects of these types of jobs are returned to later.

Likewise, any multiplier effect stemming from urban regeneration is likely to be very low. This is because the majority of goods purchased as a result of increased consumer spending are unlikely to be manufactured locally. The result is that money 'leaks' out of local economies. If increased consumer spending is concentrated on imported goods, this may lead to a worsening of balance of payments nationally (Turok 1992).

The apparent benefits of urban regeneration also vary with the scale at which the issue is examined. Projects of urban regeneration tend to be spatially autonomous; they are designed to regenerate specific spatially defined areas. The result of the proliferation of a large number of very similar projects of urban regeneration is that, when examined at a national scale, they are likely to be in competition with each other. While growth may occur in one area it is likely that this will be at the expense of decline in another area. When examined on a national scale, projects of urban regeneration may lead to no overall growth. If they have any national economic effect it is likely to be simply the creation of a new pattern of growth areas and areas in decline. This phenomenon is referred to as 'zero-sum growth' (Harvey 1989b).

Despite the evidence, the rhetoric of regeneration has tended to hold sway over rational planning and debate for much of the 1980s and early 1990s. Projects of urban regeneration are rarely accompanied by impact assessments and rarely are specific, effective mechanisms of income redistribution studied or put into place.

The predominant model of urban regeneration in North America and Europe in the 1980s and 1990s has been centred upon some form of property development or regeneration. This is based upon the belief that a strong link exists between property development and economic regeneration. However, a fundamental question mark hangs over the effectiveness of this link. Property is a sector of the economy that has never displayed any degree of long-term stability. It is characteristically susceptible to economic swings and fluctuations in property value. Investment in land and property is inherently highly speculative and the results far from guaranteed (Turok 1992; Imrie and Thomas 1993b). (See, for example, Case Study E, Urban Development Corporations and land deal losses, p. 66).

Employment issues

Effective and sustained recovery requires that localities do more than just create the impression of recovery. Not only must the surface appearance of an area be transformed but also so must the underlying economic characteristics that created the need for regeneration in the first place. The impacts of national or international economic decline are mediated at a local level. Their outcomes are not fixed or predetermined. Rather they depend upon the interaction of wider processes and local characteristics, be they economic, social, cultural and/or political. If these local characteristics are left untouched by regeneration then the future fortunes of places in relation to the impact of external forces are likely to remain similarly untouched. The conclusion must be that regeneration of this type is largely cosmetic.

Major problems which determine the local outcomes of external impacts include a lack of appropriate skills, training and educational qualifications combined with the lack of a sound local economy characterised by secure, skilled, well-paid, long-term employment opportunities, investment and innovation among local companies. To assess the degree to which the 'substance' of regeneration matches the appearance, it is important to consider the issues of the quality of employment, training, opportunity and investment present within local economies. How, to what extent, and in what ways are the local economy and labour force strengthened by regeneration.

One of the major claims of property-led models of urban regeneration is that they create jobs. These jobs stem directly from the construction of properties and indirectly in the running of developments once constructed. However, there are a number of characteristics of the construction industry that limit these benefits. Few construction jobs tend to be recruited from local populations; normally construction workers migrate to jobs from elsewhere (Turok 1992: 362–3). Even in cases where legislation has been introduced to try and encourage the local recruitment of construction labour it has tended to be unsuccessful, falling foul of official guidelines on discrimination (Loftman 1990). Also the nature of employment in the construction sector is such that jobs tend to be irregular and short term. The construction sector traditionally has a very poor record on training. Employment in the construction of property linked to urban regeneration, while it might alleviate short-term hardship, is unlikely to do anything to equip the population with useful or marketable long-term skills. Furthermore, job opportunities in the construction sector tend to be restricted to young males; therefore, large sections of the population in need of the benefits of regeneration are excluded (Turok 1992).

Serious doubt has also been cast over the importance of property in the relocation decisions of companies. It tends to be social rather than physical factors that are important in these decisions. Factors such as perceived quality of life appear more important considerations in the long-distance relocation decisions of companies (Duffy 1990; Turok 1992). Consequently, developing a supply of office space may not be a sufficient lure to firms to relocate from elsewhere. What is more likely to occur is firms moving shorter distances within labour markets, another form of zero-sum growth. Where relocations from outside local labour markets to newly developed properties have occurred it has tended to be predominantly lower-order functions (clerical and administrative) rather than higher-order functions (management, control and ownership) (Turok 1992; Graham and Marvin 1996). These higher-order functions tend to display a greater inertia than lower-order functions (a point discussed in Chapter 4). The result is the danger of a form of 'branch-plant' urban economy developing. This type of economy provides little indigenous momentum for future growth and causes urban economies to be vulnerable to decisions made elsewhere.

Problems have also stemmed from a lack of originality in the urban regeneration schemes undertaken by cities in the USA and Europe. They have typically consisted of the serial reproduction of a few models of urban regeneration. These models have usually consisted of some combination of hotel developments, exhibition or convention centres, retail parks, heritage sites, waterside developments, offices and luxury residential developments. The market for these developments is not infinite. Such unimaginative reproduction of physical developments runs the risk of creating market saturation and a consequent underutilisation of capacity. Such a possibility is obviously in conflict with the aims of those promoting regeneration (Bianchini *et al.* 1992: 254).

> Heightened inter-urban competition produces socially wasteful investments that contribute to rather than ameliorate the over-accumulation problem. . . . Put simply, how many successful convention centres, sports stadia, disney-worlds, and harbor-places can there be? Success is often short-lived or rendered moot by competing or alternative innovations arising elsewhere.
>
> (Harvey 1989b: 273)

A further problem stemming from the limitation of standard models of regeneration is that these developments are rarely well tuned to the specific nature of the localities within which they are located. The potential mis-match may express itself in a number of ways. Most obviously it presents a mis-match between the opportunities available for local people, their employment and training needs, and their present skills. This failure to utilise local expertise is socially wasteful and an inefficient use of human resources.

Undoubtedly projects of urban regeneration generate jobs within developments themselves. However, two qualifications should be applied to this statement. First, developments might have a negative impact, either locally or across wider areas, causing jobs to be lost elsewhere. Second, there is a question mark over the quality of jobs created in these developments, and their ability to alleviate conditions of poverty locally.

Projects of urban regeneration once opened may be in competition with other facilities locally. The result of this is that their success may be at the expense of job losses elsewhere in the local area. This has been most apparent where retail developments have been used as the basis of urban regeneration. The impacts of massive regional shopping developments – such as Merry Hill, near Dudley in the West Midlands, and the Metro Centre in Gateshead near Newcastle upon Tyne – on nearby older retail environments has been severe. Rather than bringing increased wealth and prosperity to regions, they may initiate new patterns of uneven economic development as well.

The patterns of employment created in urban redevelopments tend to be highly polarised. They have been characterised by a relatively small number of highly paid managerial jobs and a larger number of much less well paid, unskilled jobs in sectors such as catering, security and cleaning. This pattern of employment has led to what has been referred to as a bifurcation of opportunity (Short 1989). There exists a clear qualitative and quantitative mis-match between the few highly paid managerial jobs available in urban regeneration schemes and the needs and skills of local poorer populations. Consequently these jobs tend to go to well-qualified outsiders. Local populations tend to find their opportunities restricted to the less well paid, insecure sectors of the urban economy.

Bearing this bifurcation of opportunity in mind, it is apparent that there are serious doubts over the quality of opportunity available to poorer urban populations, those, in theory at least, with the greatest need. The jobs available to disadvantaged residents are typically unskilled, low paid, temporary or short term, non-unionised and offering poor quality training (Loftman 1990).

Local authorities have not been blind to these criticisms. They have often tried to overcome them by introducing legislation that targets the disadvantaged populations and links jobs to training programmes. However, in practice these measures have yielded disappointing results. The number of people who have benefited from them has been very low on the whole. Despite local authorities' good intentions this legislation has largely failed to overcome the limitations of the employment opportunities linked to projects of urban regeneration.

Displacement issues

Urban regeneration often involves the redevelopment of extensive areas of land. This has a considerable impact on the existing occupiers of these and other sites nearby, be they industrial, commercial or residential. There are many documented cases of the severe impact of urban regeneration on existing businesses. This is especially the case where these businesses have been regarded as incompatible to the image being imposed upon an area by regeneration. This impact is most often manifest in displacement pressures. Such pressures may stem from clean-up programmes designed to enhance the image and appearance of redevelopment areas. Displacement may have a number of adverse effects on the businesses affected. These effects include the break-up of local business networks and connections with other businesses, customers, suppliers and markets, a lack of appropriate property available elsewhere and an overall poor location. These impacts have been severe enough in many cases to lead to the closure of displaced businesses (Imrie *et al.* 1995: 34–5).

Urban regeneration also has a displacement potential on residential populations. This is examined in more detail under the heading 'Housing issues' (pp. 148–50).

Subsidy issues

Urban regeneration in British cities in the 1980s and 1990s has been characterised by partnerships between public and private sectors.

In effect this means that private sector development has been heavily subsidised by public money. This subsidy might involve a direct financial grant to private developers, for example local authorities undertaking some of the construction costs of developing a major international hotel, local authorities developing facilities, such as convention centres, which are likely to be used primarily by the private sector, or relaxing local rates or taxation to encourage business relocation. This public sector subsidy has come at the same time as a severe restriction of money to local authorities both through central government grant and money raised by local authorities through local taxation (Goodwin 1992). The subsidy of private sector development, therefore, may impose both a severe short-term cost on local authorities and a more long-term one as they pay back loans over periods of up to twenty-five years (Loftman 1990). This has resulted in some cases in local authorities raising taxes from local residents, such as the poll tax or the community charge, to subsidise a relaxation of costs on the private sector. This diversion of public funds has had far-reaching implications for the management of urban areas.

The best documented problem stemming from the diversion of public funds has been the effect this has had on social spending by local authorities. Sectors such as housing and education have often been characterised by under-funding in large urban areas in the 1980s and 1990s. This has been particularly severe in disadvantaged inner-city locations (see Case Study M.). These costs have tended to be borne by the more disadvantaged populations who typically have a higher dependency on the public sector provision of housing, health and education.

Case Study M

Diversion of public funds in Birmingham

The physical state of many of Birmingham's schools, particularly in its inner areas, was severely criticised in a report, *Aiming High*, published in 1993 by Birmingham Education Commission. The report concluded that the city had spent £250 million less than it should have done on education between 1988 and 1993, and it required £200 million to solve the problem. The report pointed out that during the same period the city had spent millions of pounds on prestigious international facilities such as the International Convention Centre.

Source: Lepkowski (1993)

The nature of the types of urban regeneration projects that these moneys have been invested in are inherently speculative. There is no guarantee, therefore, of their success, or that the public money spent will be returned (see, for example, Wynn Davies 1992). Some ventures, for example convention centres, are not designed to make money themselves. They act as 'loss-leaders' encouraging increased expenditure by visitors in other sectors of the urban economy. In effect the public–private partnership that has characterised urban regeneration in the UK involves the public subsidy of private profit (Harvey 1989b). Further to this, a number of critics have questioned the need actually to subsidise private investment. They argue that in many cases private investment does not necessarily respond to public subsidy (Turok 1992: 374–6).

Problems with the type of investment encouraged

The type of investment targeted by urban regeneration projects is inherently unstable. It is likely to fluctuate with swings in the economy. This is the case, for example, with the business-tourist market. Such investment is also geographically footloose and is prone to rapid switches of location in response to minute variations in the social characteristics of locations.

The retail sector has formed an important part of many urban regeneration developments. These have included regional shopping 'mega-malls', such as the Metro Centre in Gateshead, and festival shopping developments, such as those in the converted warehouses of London's Docklands. However, certain characteristics of the retail sector severely limit its effectiveness as a promoter of sustainable, fair economic regeneration. Employment opportunities in the retail sector open to disadvantaged populations are limited. They are typically of a low quality, fitting the profile of poor employment opportunities outlined earlier in this chapter. Encouraging retail developments also potentially engenders problems for consumers. The retail boom of the late 1980s was fuelled largely through a boom in the availability of personal credit. This later led to many problems of personal debt as consumers struggled to repay the amounts they had borrowed plus the associated interest (Turok 1992).

Summary: economic issues

First, urban regeneration projects engender a highly uneven distribution of costs and benefits. The benefits tend to accrue to a small number of professionals from the business and managerial classes. These are often the people who would have done well anyway, without regeneration. The groups in most need of economic benefit, the disadvantaged populations occupying deindustrialised inner-city areas or run-down peripheral housing estates, tend instead to bear the brunt of the costs and negative impacts of urban regeneration (see Case Study N).

Social issues

The dual city?

The bifurcation of economic opportunity described in the previous section has had an inevitable influence upon the social geography of cities (Short 1989). A metaphor frequently used to describe these new social geographies is that of the 'dual city'. The idea of a dual city is based upon evidence of increasing social division within cities and the apparent emergence of an urban 'underclass' divorced from dynamic mechanisms in the formal economy. This economic exclusion is translated into exclusion from many areas of city life. This underclass consists of both waged and unwaged poor, a disproportionaly high number of members of ethnic minorities and groups such as sick, elderly or disabled people and single parents. It has been argued that as social polarisation has increased as both a result of international economic trends and recent government policy. This social polarisation has had an imprint on the spatial structure of cities. Areas such as inner-city areas and peripheral council housing estates have become equated with the presence of this urban underclass. These pockets of deprivation contain a number of groups who are surplus to the requirements of the formal urban economy. These areas are characterised by the prevalence of 'alternative' or 'twilight' economies run on an informal or illegal basis.

These pockets of deprivation have become contrasted to highly spectacular 'islands' of regeneration which exist often in very close proximity to these 'seas of despair' (Hudson 1989). London's Docklands provides an example of just such a spectacular island of regeneration set among some of the poorest communities in London. In Birmingham the

Case Study N

Urban renaissance: myth or reality in Cleveland, Ohio?

During the 1960s, 1970s and much of the 1980s, Cleveland, Ohio, was regarded as the epitome of all that was wrong with urban America. It was polluted, so much so that its river actually caught fire in June 1969, its economy was bankrupt, its leaders were widely regarded as incompetent and provincial and its baseball team had not graced the World Series for over thirty years. The city's epithet said it all: 'the mistake by the lake'. However, by the 1990s the talk of Cleveland was of a city reborn. The most tangible symbol of the alleged rebirth was its landscape. This included a $200 million stadium for the now successful baseball team, a post-modern showcase development housing the Rock and Roll Hall of Fame, as well as a host of other impressive architectural developments. What is more, herons can land on the Cuyahoga river without fear of getting their feathers singed.

Cleveland's apparent transformation dates back to the conversion of the terminal of the Baltimore and Ohio Railway into a leisure, retail and commercial development in the early 1980s. The process has since involved the business community, non-profit foundations and new municipal leaders.

Cleveland's renaissance has not passed without criticism, however. Clearly the renewal projects have been undertaken with particular audiences in mind. These include middle-class suburban Cleveland rather than its poorer inner-city. For example, while the city's downtown has been extensively revamped the same could not be said of its school system (Cornwell 1995).

While Cleveland looks like a city on the up, the extent to which all in Cleveland could be said to be sharing this trajectory is more questionable. Cleveland is an example of an American 'urban success story', a city trumpeted by the media, marketers and politicians as a successfully revitalised city. A number of other cities in the USA and Europe have made similar claims. However, such claims are typically light on objective, or tangible measures of regeneration. The question that remains is: how real is this regeneration? The question of what is meant by real is clearly problematic and unlikely to draw universal agreement. However, a basic requirement of a claim to meaningful, 'real' regeneration might be that the economic well-being of residents has improved. While this is clearly not the only measure that could be employed, it appears to be a fundamental one.

To evaluate the claims of these urban success stories for North American cities it was necessary to draw up a list of cities that were 'distressed' in 1980. This was done using an index which included measures of unemployment, poverty, household income and income change, and population change. From this a list of fifty economically 'distressed' cities was compiled. It was necessary next to find out which of these cities had been held up and widely regarded as a success story since 1980. To discover this a range of expert opinion was polled; these experts included academics and economic development practitioners. From their responses two classes of 'successful' cities were drawn up: twelve 'successfully revitalised' cities, mentioned by 20 per cent or more of respondents and of these six 'most successfully revitalised' cities, mentioned by 40 per cent of more of respondents (Table 8.1). These groups were based on the respondents' perceptions rather than any objective measures of revitalisation. It was important to assess the extent to which these perceptions accorded with more objective measures of revitalisation.

To evaluate the claims of these success stories the 'successfully revitalised' cities were compared to those cities deemed 'unsuccessful'. An index of the performance of these cities between 1980 and 1990 was constructed using measures of residents' economic well-being.

The results of this evaluation were revealing. They showed that there was no significant difference in the performance of 'successfully revitalised' cities compared to 'unsuccessful' cities. Stories of urban revitalisation and renaissance appeared not to translate into tangible improvements in the economic well-being of the residents of these cities, compared to those of 'unsuccessful' cities. In fact many 'unsuccessful' cities performed better than 'successful' ones over the period on a majority of indicators. Of the six 'most successfully revitalised' cities, only Atlanta and Baltimore significantly outperformed the

Table 8.1 *'Successfully revitalised' cities*

Central city	State	Count	Response (%)
Pittsburgh	Pennsylvania	63	82.9
Baltimore	Maryland	49	64.5
Atlanta	Georgia	40	52.6
Cleveland	Ohio	37	48.7
Cincinnati	Ohio	33	43.4
Louisville	Kentucky	31	40.8
Miami	Florida	23	30.3
Boston	Massachusetts	22	28.9
Chicago	Illinois	22	28.9
Birmingham	Alabama	18	23.7
Buffalo	New York	18	23.7
Norfolk	Virginia	16	21.1

Source: Wolman *et al.* (1994: 838)

'unsuccessful' cities; of the 'successfully revitalised' cities, only Boston achieved this.

Cleveland's performance in this evaluation paints a rather different picture from that of the press report discussed above. Cleveland performed significantly more poorly than the average of 'unsuccessful' cities on all indicators used: percentage point change in unemployment, percentage change in median household income, percentage change in person's below the poverty line, percentage change in per capita income, and percentage point change in labour participation rate.

Two conclusions can be drawn from this evaluation of 'urban success stories'. First, such success stories appear to be largely based on image and perception rather than objective measures of well-being. Second, those cities heralded as urban success stories tend to be those who have displayed an 'appearance' of revitalisation. Namely, those cities who have physically upgraded their central areas and enhanced their images. However, there is no necessary correspondence between an appearance of revitalisation and its realisation in terms of residents economic well-being. Cleveland's claims to regeneration appear particularly hollow in this light.

Sources: Cornwell (1995); Wolman *et al.* (1994)

International Convention Centre built at a cost of £180 million is only 200 yards from Ladywood, one of the city's poorest wards (Loftman and Nevin 1992). While the picture painted here is something of a generalisation and the unique characteristics of individual cities should not be ignored, it does highlight an important trend in the social geographies of urban areas in Britain and to a greater degree in the USA.

Urban regeneration and social policy

Social improvement appears to be the indirect, rather than the direct, consequence of urban regeneration projects. The social and community policy aspects of urban regeneration programmes have typically received a very low proportion of total regeneration spending (Table 8.2). They also tend to be the most vulnerable aspects of these programmes during times of economic recession.

There is little evidence that those outside the formal job market have been included in the social regeneration policy aspects of urban regeneration programmes. Social regeneration has been all too frequently equated with physical and economic regeneration.

Table 8.2 *Urban Development Corporation spending on community projects (% of total annual spending)*

	1988–9	*1989–90*	*1990–1*
Black Country	1.3	1.8	2.7
Bristol	0.0	1.0	0.0
Cardiff Bay	0.6	0.9	1.0
Central Manchester	1.4	1.6	3.4
Leeds	0.9	0.1	0.2
London Docklands	3.3	6.5	5.3
Merseyside	4.5	3.1	4.2
Sheffield	0.4	0.7	1.0
Teesside	2.4	0.3	0.4
Trafford Park	0.1	0.3	1.4
Tyne & Wear	0.3	0.9	1.3
All	1.38	1.56	3.70

Source: Imrie and Thomas (1993:16)

Housing issues

Urban regeneration projects affect the local housing stock in two ways. These are physical displacement of poorer resident populations and displacement through forcing house price increases. Physical displacement may occur where residential properties are demolished to make way for newer developments. It is, inevitably, cheaper, low-value accommodation that is removed and which serves low-income, marginalised populations. In some cases hotel and rooming accommodation which served as a valuable source of cheap accommodation has been renovated and modernised to cater for new, often tourist, markets which are expected to flood into areas following regeneration. The consequences of this have been that poorer residents have been evicted in the short term and have suffered a drastic reduction in the supply of affordable accommodation available to them in the longer term. The physical and psychological impacts of this have been severe on residents who may have been elderly or suffering from long-term illnesses. Frequently there has been no council housing provision to compensate for displacement. Displacements such as these have been common in the large-scale redevelopment of North American cities, for example in association with the development of Vancouver's Expo '86 site (Beazley *et al.* 1995).

Low-income populations may also suffer displacement through being squeezed out of local housing markets as regeneration forces the value and hence the price of local housing up and out of the range of low-income populations. Areas around regeneration projects have frequently been subject to an increase in property speculation and interest from property developers with an eye on encouraging the influx of middle-class professionals with a taste for city centre living. This interest adversely affects the ability of low-income populations to purchase property locally, forcing them into alternative cheaper property markets, often in different localities. This is an example of a process known as gentrification. Although by no means caused exclusively by urban regeneration, it may be exacerbated by it. This has proved a greater problem than direct displacement in British cities to date. Other development pressures suffered by those adjacent to redevelopment projects have included the loss of community facilities to make way for redevelopment and disruption and noise population during the construction phase of development.

Flagship projects of urban regeneration frequently include programmes of housing provision or renovation. Waterfront developments typically utilise the legacy of distinctive architecture and original features, converting them into, for example, luxury housing. Again, they are aimed at luring incoming professionals. This accommodation does nothing to solve housing and accommodation crises among local poor populations who are unable to take advantage of these developments. Housing developed in association with such projects of urban regeneration tends to be inappropriate to local populations on a number of counts. First, it is far too expensive. The majority, often over 80 per cent, of this housing is owner-occupied. During the housing boom of the late 1980s some two-bedroom flats in London's Docklands were selling for £200 000, while the majority of households in the surrounding boroughs had combined annual incomes of less than £10 000. Further, the types of properties provided in these developments are aimed primarily at the 'dinky' market (double income, no 'kids', typically young professional couples in their twenties and early thirties). Consequently this accommodation primarily consisted of small flats. The accommodation needs of the majority of local households are usually very different from those of the 'dinky' couple. Often families are older and with children. They require housing with three or more bedrooms and preferably gardens and play areas. The areas surrounding inner-city regeneration schemes often contain a large proportion of families from Asian ethnic minorities. These communities are characterised by a high proportion of extended family households.

Their requirements are again for the larger housing units notoriously absent from urban regeneration schemes.

The provision of 'social' housing, affordable accommodation aimed at addressing local housing needs, has long been a major plank of the social regeneration programmes of urban regeneration schemes. In Cardiff Bay, for example, it was agreed that 25 per cent of new housing in the development should be social housing (Case Study O). However, the actual delivery of this social housing has tended to be very disappointing. Frequently, there is little attempt to match the housing provided with the characteristics of the local population. All too often, as well, the actual amount of social housing in completed regeneration schemes is well below the amount initially agreed (Rowley 1994). Often the need to generate profit from housing ultimately overrides social considerations.

Issues of local democracy and public involvement

Urban regeneration has taken place within changing structures of urban governance. (This was discussed in detail in Chapter 5.) The creation of agencies, such as Urban Development Corporations and Training and Enterprise Councils in the UK, has effectively taken certain areas out of the control of local authorities and transferred accountability away from the local area and towards central government. This and the other political characteristics of urban regeneration have been accused of imposing a barrier to public involvement in decision-making and ultimately a threat to local democracy as it has developed in British cities.

Projects of urban regeneration in the UK, Europe and North America have been characterised by a lack of public debate, consultation, inquiry or detailed prior impact assessment. The exclusion of the public from the development process in this way has been facilitated by wider political and legislative frameworks which are biased in favour of 'growth-coalitions' and against the inclusion of public and opposition groups (Beazley et al. 1995). Debates, for example on the costs and impacts of urban regeneration, which, prior to the introduction of centrally appointed agencies such as Urban Development Corporations, were conducted in the public sphere, now take place as part of private board meetings (Imrie et al. 1995). Furthermore, legislation in the Local Government Act 1985 allowed information to remain confidential and away from public scrutiny (Beazley et al. 1995).

Case Study O

Housing provision and improvement in Cardiff Bay

The Cardiff Bay Development Corporation (CBDC) proposed the development of 6 000 housing units and improvement to the 2 000 existing units in their area. Three-quarters (75 per cent) of the new units were to be for private owner-occupation, with the remainder as social housing. A wide range of accommodation types was envisaged which included sheltered accommodation for elderly people, family homes, single person and starter units, and prestige waterfront developments. Three areas had been developed by 1994. Windsor Keys is a much in demand waterside location. Despite being 25 per cent social housing, residents felt much dissatisfaction with the development. They have suggested that the percentage should be raised to provide a more balanced development. The CBDC argued that the high proportion of private housing was required to fund the development of social housing. Despite the provision of social housing the CBDC made no provision to meet the needs of the large elderly population nearby. Another of the developed sites is in Tyndall Field near the city centre. Again the planning brief stated that 25 per cent of the housing should be social housing. The final developed site, Butetown, is a large area of council housing. While no sites here were recognised for development, the area underwent piecemeal refurbishment. This caused almost unanimous dissatisfaction among residents. Residents felt that investment promised for the area had not been forthcoming and that the CBDC ignored the needs of local residents for housing, employment and education. They believed that the refurbishments that had occurred were irrelevant to the needs that existed on the estates. For example, while a bottle bank had been provided, broken elevators, faulty launderettes, leaking roofs and windows remained. The CBDC had also failed to implement their suggestion that homes should be provided for Butetown's large ethnic population throughout the area.

Source: Rowley (1994)

In the face of powerful regeneration coalitions supported by legislation, community groups and opponents to regeneration appear weak, under-resourced and distanced from power. In financial terms they are unable to match the powerful regeneration coalitions: in the majority of cases these groups are funded by donation or subscription. Occasionally they are funded by grants from private companies or local authorities. However, this has been a source of controversy as these are often the very organisations that are being opposed. Opposition groups are further excluded from debate by the restricted and limited information divulged by regeneration coalitions, a lack of legal or technical support afforded to opposition groups in comparison to that available to regeneration coalitions and the increasingly 'closed' nature of decision-making in local government. When faced by apparently unified regeneration coalitions, opposition groups appear fragmented because of their diversity of agendas. Consequently public involvement in regeneration programmes has been extremely limited. While local authorities and regeneration coalitions might promote the appearance of public involvement, through, for example, public meetings, these are often well after development has begun and they have no influence on decision making or policy (Beazley *et al.* 1995). Recent urban policy, however, particularly in the USA, appears to be more democratic, featuring community representation and some degree of local accountability.

Cultural issues

Culture has been an important component of the transformation of both the economies and the images of cities during the 1980s and 1990s. The cultures of cities have been affected in several important ways by their implication in the process of urban regeneration. 'Culture' includes not only cultural activities (entertainment, theatre, opera, ballet, music) associated with economic or social elites, but also those activities that constitute the 'everyday' or the 'ordinary', a more democratic notion of culture. These two meanings are often distinguished by the prefixes 'high' and 'popular'.

The cultural city

The development of new cultural facilities within cities has been led primarily by the need to appeal to wealthy external audiences and to complement urban regeneration developments like prestigious hotels and convention centres. These facilities are typically very exclusive; they hold little appeal to the majority of the urban population either because of the activities that go on there or because of the expense involved. They may also be difficult to reach for populations with a heavy reliance on public transport, who once there might find the atmosphere unwelcoming. Further, it is likely that the programmes on offer will be of little appeal to certain groups, particularly to ethnic or cultural minorities whose interests are usually very poorly provided for in international flagship developments (Bianchini *et al.* 1992).

The development of prestigious cultural facilities such as international concert halls is actually more likely to narrow the cultural profile of cities. This is so for two reasons. First, in most cases the development of prestigious cultural facilities goes hand-in-hand with attempts to attract major arts companies (ballet companies or orchestras for example), hosting cultural festivals of literature, drama, music, film, television and sport. A consequence of this is that attention within cities is shifted towards high culture of international standing and with international appeal. Less prestigious community-based cultural activities may suffer from cuts in funding forcing them to restrict their activities or to close altogether. The brunt of these cuts fall on minority or community arts, especially of ethnic minorities (Lister 1991).

The use of culture within economic development has, on occasions, extended to the appropriation of formerly community-based and led activities, by local authorities, and a subsequent incorporation into 'official' cultural programmes. The result may be that community groups lose control over activities that were formerly their own (Case Study P).

Excluded cultures and protest

The promotion of prestigious cultural facilities and activities of international appeal as part of programmes of urban regeneration has led to accusations from groups representing minority and community-based cultures that their contribution to the cultural life of cities is undervalued. Opposition from groups such as Workers' City, a group promoting the working-class history of Glasgow, to the European City of Culture celebrations mirror a feeling that is widespread in cities undergoing similar 'make-overs' (see Case Study Q). These groups, which typically represent a diversity of cultures including working-class communities, ethnic minorities, women's organisations and gay and lesbian communities, have commonly felt that their culture has become devalued because it does not represent a 'marketable' commodity on the international circuit. This has prompted resistance by these groups to what they see as the shallow or 'facsimile' cultures promoted by urban regeneration. Such groups argue that they represent more 'genuine', 'organic' or deeply rooted cultures than those being promoted.

The means by which their opposition to the cultures promoted by urban regeneration is manifest are varied. Marches and demonstrations have long been a prominent form of urban protest. The history of the

---

Case study P

Hijacking the Handsworth carnival, Birmingham

The Handsworth carnival in Birmingham was first held in September 1984 to provide entertainment for the largely black population of the city's inner area. It attracted 50 000 people in its first year. By 1991 this had risen to 500 000. In the early 1990s the carnival attracted considerable financial backing from Birmingham City Council. This raised fears among black residents that the carnival might be appropriated into attempts by the City Council to put the city on the international map. In 1991 the local organising committee was wound up and the carnival was renamed the 'Birmingham International Carnival'. Previously the Council had distanced its association with the carnival in the Handsworth area. However, once the City Council began to promote the city internationally it was able to generate good publicity by demonstrating that it was supporting a black community initiative. This added weight to Birmingham's claim to be 'an international city'. At the time speculation was rife that the carnival was to be moved to a location within the city with a less problematic image, leading to fears that the carnival would lose touch with the very communities that it was originally designed to help. There were worries that the involvement of the City Council would lead to the carnival becoming primarily a tourist attraction, sanitising black arts and culture and attracting a predominantly white outside audience.

Source: Taylor (1991)

---

development of London's Docklands development from the early 1980s has been dogged by protest from community groups such as The Docklands Forum. Such demonstrations have included the 'People's Armada' in 1984 where a number of boats supporting community protests sailed past the Houses of Parliament (G. Rose 1992). Frequently opposition is expressed by less direct means and includes the articulation of community, history and a sense of belonging to a place threatened by development. Such protests have included publishing oppositional accounts of places often incorporating working-class social histories, poster and arts campaigns, newsletters and petitioning local and national authorities (J.M. Jacobs 1992; T. Hall and Hubbard 1996).

The nature and organisation of oppositional groups have also been very diverse. Some are well organised and funded and focus their intervention around specialist issues such as democratic planning and development, drawing on the expertise of researchers and professionals, for example, the community planning group Birmingham for People. Other groups employ the talents of artists to help articulate community feeling. A good example of this was the Docklands Community Poster Campaign involving the Art of Change group (Dunn and Leeson 1993; see also Plate 5.1). Other groups may be much more informal and spontaneous, forming to protest against specific developments (Case Study Q). Some of the most frequently opposed aspects of urban regeneration developments include the physical, economic, cultural and psychological impacts of urban regeneration on communities, the lack of public consultation on development and the bypassing of existing planning legislation, and the lack of social and communal facilities in new developments.

Case Study Q

Opposition to the European City of Culture in Glasgow 1990

Workers' City is a radical organisation consisting, among others, of writers and militants who vehemently opposed Glasgow's role as European City of Culture in 1990. The basis of their opposition was that the version of Glasgow's culture promoted through the year of celebrations failed to reflect what they felt was the 'real' cultural experience of the city, the working-class experience. Rather, they argued that the celebrations promoted only a shallow or 'facsimile' version of culture that was designed to appeal to international visitors to the city, rather than its own working-class residents. The celebrations imposed a 'glitzy' image on the city and failed to provide any attractions appropriate to Glasgow's working-class population. They also imposed a considerable financial burden on the local authorities at a time when funds were desperately needed for the physical regeneration of the city's working-

class housing stock, particularly tenement buildings and peripheral housing estates. The celebrations included some 4 000 events and nearly 10 000 performances which cost approximately £50 million. Of this, Glasgow District Council contributed £15 million from a special fund and Strathclyde Regional Council contributed £12 million. The local economy undoubtedly benefited from the celebrations, especially the spending of an estimated 4 million visitors to the city during 1990. However, Workers' City point to the advertising budget of £2 million paid to Saatchi and Saatchi and the £100 000 paid in consultancy fees to the architect Doug Clelland to renovate some arches below the Central railway station as evidence of the celebrations clouding the spending priorities of the local authorities.

Source: Cusick (1990)

Assessing the impact and influence of such groups is difficult, given their heterogeneity and the diversity of their aims and approaches. Their involvement with local authorities and developers can range from close consultation and involvement to outright hostility and confrontation. While it is dangerous to overgeneralise, it is probably true to say that the impact of these groups has been relatively marginal to the aims of local authorities and developers. Often the concessions granted to protesters have been token or developers have appeared to take note of community issues largely to give the appearance of democracy, consultation and communal involvement in development (T. Hall and Hubbard 1996: 166).

Future urban regeneration

Attitudes towards the regeneration of urban areas in the UK, Europe and North America appear to be becoming more circumspect. This is a result of the criticisms that regeneration schemes initiated in the 1980s and

---

Case Study R

The Earth Centre, Doncaster: a green flagship?

The Earth Centre is an unusual attempt to regenerate an area physically and economically scarred by the closure of two coal-mines in 1979 and 1986. Located on the site of the once prosperous South Yorkshire coalfield, the Earth Centre aims to regenerate environmentally a 350-acre site and provide an education and entertainment facility on the theme of the future health of the planet. It was founded by Jonathan Smailes, a former director of Greenpeace, in 1990 and is due to be completed by the year 2 000. The Centre will contain an organic farm, a museum, a 'clean' factory producing household goods, a 'green' hotel, entertainment facilities and new buildings designed by the progressive architects Future Systems that will demonstrate energy efficiency and non-polluting technologies. The Centre aims to be self-sustaining ecologically and financially. The Centre also hopes to contribute to the economic regeneration of the blighted South Yorkshire area. It has already utilised voluntary labour from school children and former miners and hopes to employ far more people as the site nears completion. The Earth Centre is a model of regeneration that aims to provide substance, pride and environmentally positive impacts in contrast to the theme parks and shopping mall projects operating nearby.

Source: Glancy (1995; 1997)

early 1990s received from academics, researchers and the media, the persistence and even worsening of several major urban economic and social problems and the continued financial crises faced by local authorities. The image of the property-led model of urban regeneration has been severely tarnished by a number of high-profile failures and disasters. These have included the bankruptcy of Olympia and York in 1992, who were the developers behind London Docklands' Canary Wharf. Given this recent history it would seem likely that future developments might pay greater attention to issues such as social welfare, economic and environmental sustainability and the ethnic and cultural diversity of urban populations. This attitude is reflected in recent urban policy, particularly from the USA (see Chapter 5). It is possible that, rather than the monotonous imitation of models of urban regeneration by city after city, the future will see projects more attuned to the characteristics of their localities or which utilise local pools of skills (Bianchini *et al.* 1992), or even 'green' flagship developments reflecting issues of global concern such as the environmental crisis (Case Study R).

Further reading

There are few books that really explore the impact of urban regeneration projects in any great detail. However, a number of useful case studies appear in:

Hall, T. and Hubbard, P. (eds) (1998) *The Entrepreneurial City: Geographies of Politics, Regime and Representation* Chichester: Wiley.

Healey, P., Davoudi, S., O'Toole, M., Tavsanoglu, S. and Usher, D. (eds) (1992) *Rebuilding the City* London: Spon.

Imrie, R. and Thomas, H. (eds) (1993) *British Urban Policy and the Urban Development Corporations* London: Paul Chapman.

The theme of urban regeneration and the fashioning of unequal cities is taken up in:

Ambrose, P. (1994) *Urban Process and Power* London: Routledge.

Jacobs, B.D. (1992) *Fractured Cities: Capitalism, Community and Empowerment in Britain and America* London: Routledge.

Good sources of information are the regular reports on urban regeneration projects which appear in journals such as *Environment and Planning*, *Planning Practice and Research*, *Town and Country Planning* and *Urban Studies*.

⑨ Urban futures

- Scenarios for future cities
- Sustainability and urbanisation – meanings and prospects

The sustainable city: the next utopia?

Sustainable urban development is a phrase that generally describes a form of development that does not create negative environmental impacts. Sustainable urban development literally refers to a type of urban development that could be maintained, in theory, indefinitely without destroying the resources upon which it depends (Harris 1995). The general sharpening of environmental awareness world-wide has produced a series of debates on sustainability which have focused upon the issue of how existing urban forms might be managed with less severe environmental impacts and how future urban developments might be made environmentally benign. Cities are the major consumers of the world's non-renewable energy resources; they are also the world's major producers of pollution and waste. The city is also (and is likely to continue to be) the locus of both major population migration and population growth. Much of the current environmental 'crisis' is seen as either directly or indirectly attributable to cities.

However, sustainability has come to mean something far wider than just environmental sustainability. Sustainability has become a byword to describe an equitable form of urban development which encompasses social, economic, political, cultural and moral sustainability as well as environmental sustainability. How these measures might be achieved has been the subject of intense ongoing debate and speculation between academics from a variety of the natural and social sciences, who, despite their differences, are united in their belief of the desirability of sustainable urban development.

Utopianist visions of future urban development have been a common characteristic of urban thought and writing as long as there have been cities. They have been especially common during times of urban crisis, or during the perception of urban crisis. For this reason it is not surprising to see such intense speculation over the nature of urban development now. However, few common paths or policies have been successfully or widely adopted, few barriers, political, economic or cultural, to a successful, sustainable urban future have been dismantled. It is apt then to consider how likely the vision of a sustainable urban future is to be realised. This chapter will briefly reconsider the five major processes that are likely to shape urban development into the early twenty-first century. To what extent and in what ways will these processes help or hinder the hoped-for procession towards an environmentally, economically, socially, politically, culturally and morally sustainable urban future?

Five scenarios for sustainability

The global city

It is very apparent that the world economy is becoming increasingly interconnected and that major global cities are the nodes through which many of these connections run (Hamnett 1995; Knox 1995). Some consequences for these cities include the reorientation of their economies around financial and producer services (Sassen 1991; 1994). The global status of these cities is made manifest in distinctive, spectacular landscapes. However, on current evidence it would appear unlikely that a global world economy will produce cities that demonstrate any great degree of economic or social sustainability.

Despite being able to generate superprofits, the instability inherent in the world's financial system means that global cities are also liable to bear the consequences of occasional 'super-losses'. The London Stock Market collapse in 1987 is an example of such a super-loss. This crash almost instantaneously wiped billions of pounds off the values of shares. Such instability precludes the formation of any stable base upon which to build any long-term economically sustainable system. This inherent instability is exacerbated by the susceptibility of major financial centres to the effects of massive financial fraud. The collapse and near collapse of a number of prominent international banks and financial institutions following fraud highlights this danger. The global nature of financial networks ensures that the impacts of these frauds are felt in cities and by companies and individuals world-wide. Urban geographers are beginning to realise that financial 'scams' are a major factor in the urban geography of global financial centres (Soja 1995). The employment patterns that are created in global cities by the development of international financial economies also tend to be very polarised. This creates highly unequal divisions of income and opportunity in these cities.

Finally, the increasingly networked world economic system has opened up a very stark division between those cities who are integrated into this 'fast-world' and those who are not, and who are relegated to the 'slow-world' (Hamnett 1995; Knox 1995). Regions in the slow-world include declining former industrial regions and rural areas with quantitatively and qualitatively diminishing populations. These regions suffer from 'peripheralisation'. They become progressively, and apparently irreversibly, distanced from the opportunities available to those centres linked to the world economic network. They become receivers of economic fortune from elsewhere rather than instigators of economic development.

These factors would suggest that the development of global cities and an integrated world system appears likely to produce patterns of social, economic and spatial inequality within and between cities.

The competitive city

Cities have, for a number of reasons, had to become more competitive in the pursuit of investment and jobs. This competitiveness has involved policies and programmes which have had the intention of encouraging economic development, speculative urban regeneration schemes and place promotion. Within a competitive city, and this competitive condition now appears almost ubiquitous, the priority is generally economic development rather than sustainability. The efforts employed to this end tend to be massive consumers of economic and social resources. A flagship urban development might easily cost in excess of £100 million. Such ventures are usually speculative and financed heavily from the public purse. Some consequences of these developments include a reduction in spending on health, social welfare and education, saddling local authorities with long-term debt and interest repayments and vulnerability to the failure of regeneration projects. The almost universal adoption of a few models of regeneration has led to decreasing returns on investment and wasteful reproduction of facilities. Furthermore, such programmes have appeared to produce a series of negative impacts which fall disproportionately on those groups least able to afford them. Competitiveness within cities does not appear to foster an environment in which genuinely sustainable notions of development can be easily explored.

The electronic city

The development of telecommunications technology has, and will continue to, affect the development of the physical form of cities. How exactly this will reorient urban geographies is, as yet, unclear. Telecommunications have had contrasting effects on urban development. They have encouraged the decentralisation of activities away from city centres. Spatially dispersed networks of factories, offices and research and development units can now be linked by sophisticated communications networks. The availability and the sophistication of these systems is only likely to increase in the future. The development of this decentralisation has the potential to enhance sustainable urban development. The decentralisation of employment and the development of telecommuting is likely to reduce the volume of centre–suburb commuting, make journeys more flexible with regard to time, and ease the burden on the urban transport infrastructure during rush hours. Also, sophisticated systems of traffic monitoring which employ telecommunications technology, although currently in their infancy, are likely to make using the transport infrastructure of cities more efficient (Graham and Marvin 1996: 327–33).

However, this potential needs to be balanced by the possible negative impacts of telecommunications on cities. The development of networks of telecommunications are unlikely to proceed freely and evenly. They are likely to be structured by existing social and economic dynamics. The development of these networks, therefore, will probably contribute towards enhancing existing social and economic inequalities. The consequences of this are likely to include the social and spatial concentration of telecommunications facilities in, for example, the centres of large, global cities. These network nodes, and the people and institutions who control access to them, will probably become increasingly powerful within the electronic economy and society. The corollary of this is likely to be the creation of spaces and social groups excluded from electronic networks, 'electronic ghettos' peopled by an 'information underclass'. The development of an increasingly electronic society is unlikely to be free of the forces that have created inequalities within the existing society.

The edge-city

It has been argued that a decentralised urban form characterised by edge-cities (city-like settlements on the fringes of existing urban settlements) offers an environmentally sustainable form of urban development. This argument cites the decrease in existing commuting patterns as evidence. However, this has been countered by the argument that small, compact cities around a single centre offer the most efficient and sustainable urban form. Whether or not the edge-city is able to contribute to environmental sustainability is still a matter of some debate. However, what is becoming clear is that it fails to offer a model of urban development that appears able to enhance social sustainability. Edge-cities, in the forms that they have appeared in the USA to date, have been socially very exclusive. They have largely taken the form of large, private, master-planned communities aimed exclusively at affluent consumers. Little or no provision is usually made for other, less well-off, social groups. Indeed the exclusive nature of these settlements has been emphasised by the fact that plans to provide affordable housing, for example for teachers, have been frequently met by protest (Knox 1992b). The development of the exclusive edge-cities around existing cities is also likely to have a detrimental effect on its wider urban setting. As the affluent retreat behind gated suburbs other parts of the city are likely to become progressively abandoned and downgraded as social and economic prospects worsen (Beckett 1994: 12). Rather than social sustainability edge-cities seem to represent a form of urbanisation that contributes to an increase in social, economic and spatial sustainability.

The creative city

It is from cities, or rather creative and innovative individuals, communities and areas within cities, that the forces that have shaped economic, cultural, political and artistic life have emerged throughout history. The capacity for cities to unearth invention and dictate such change has been dubbed creative capacity. It has been argued that creative and innovative ways of negotiating and shaping the severe structural changes affecting cities have been developed within the spaces of decay in otherwise declining cities during the 1980s and 1990s (P. Hall 1995; Landry and Bianchini 1995). The attention focused on the spectacular regeneration of city centres during the 1980s has tended to divert attention away from these less spectacular but potentially more sustainable developments.

It is difficult to classify or fully describe the range of ways in which this creative capacity has been manifest in the solution to urban and economic problems. However, a characteristic of the current wave of urban creativity is the innovative utilisation and combination of skills derived from artists, designers, educators and entrepreneurs with new technologies such as the Internet. At the moment this creative capacity is expressing itself in a number of small-scale, locally oriented mileux within European cities. These creative communities have broken down barriers that have existed between, for example art and technology, work and leisure in the solution to the problems they have found themselves facing. Examples of creativity in an urban context are myriad and diverse. They include ethnic business networks, cultural quarters, the imaginative transformation of public space, projects to marry art, urban design and technology and many more. Indeed the failure to precisely define exactly what is meant by creativity in an urban context is the main thing limiting understanding of the process at the moment.

Since the failure of property-led models of urban regeneration have come to light in the 1990s, academics, the media, entrepreneurs and policy-makers have begun to recognise the potential of the responses to structural change by groups previously seen as marginal to the solution to urban problems. These groups include artists, small businesses, ethnic and cultural minorities and young entrepreneurs. As cities approach the millennium the diverse creative capacities of their populations are increasingly being recognised as important assets shaping urban futures.

The ways in which this creative capacity is nurtured, realised and utilised by communities, corporations and those responsible for city government at local, national and international levels and the tensions and relationships between creativity and the evolving imperatives of capitalism are likely to form a significant strand of urban geography well into the twenty-first century.

Conclusion

The scenarios sketched out paint a fairly gloomy picture of the likely paths of future urbanisation. Although they all reflect powerful processes of urbanisation operating in cities at the moment, none of the above scenarios is inevitable. Urban development will not simply reflect one or other of the models outlined; rather, it will be different in every city. It is likely to reflect some combination of each of the above models mediated through local circumstance. These models of future cities will not be hammered out across space without resistance. Urbanisation will probably reflect a number of other, sometimes contrary, forces to those outlined. There are also many examples of innovative and progressive forces of urbanisation at work in the city. The battle for ascendancy will take place at the local level as well as the global and it is likely that many alternative cities will continue to be found in the cities of the future, as they are in the cities of the 1990s.

Further reading

Readings on the global city can be found at the end of Chapter 3. Readings for the competitive city include those recommended at the end of Chapter 4 plus the following article:

Hall, T. and Hubbard, P. (1996) 'The entrepreneurial city: new urban politics, new urban geographies?' *Progress in Human Geography* 20, 2: 153–74.

The most comprehensive guide to the electronic city is:

Graham, S. and Marvin, S. (1996) *Telecommunications and the City: Electronic Spaces, Urban Places* London: Routledge.

Readings on the edge-city are to be found in:

Platt, G. and Macinko, G. (eds) (1983) *Beyond the Urban Fringe* Minneapolis: Minneapolis University Press.

The best guide to sustainable urban development is:

Haughton, G. and Hunter, C. (1994) *Sustainable Cities* London: Jessica Kingsley.

Scenarios for future urban development are explored in the following series:

Brotchie, J., Batty, M., Hall, P. and Newton, P. (eds) (1991) *Cities of the Twenty-First Century* Melbourne: Longman Australia.

Brotchie, J., Batty, M., Blakely, E., Hall, P. and Newton, P. (eds) (1995) *Cities in Competition* Melbourne: Longman Australia.

A good discussion of creativity and urban change is:

Landry, C. and Bianchini, F. (1995) *The Creative City* London: Comedia/Demos.

References

Allen, J. (1988) 'Towards a post-industrial economy?' in Allen, J. and Massey, D. (eds) *The Economy in Question* London: Sage.

Allen, J. (1995) 'Crossing borders: footloose multinationals' in Allen, J. and Hamnett, C. (eds) *A Shrinking World? Global Unevenness and Inequality* Oxford: Oxford University Press.

Ambrose, P. (1994) *Urban Processes and Power* London: Routledge.

Arlidge, J. (1994) 'Glasgow's hopes shrivel after the hype' *The Independent on Sunday* (2/10/94): 4.

Ashworth, W. (1968) *The Genius of Modern British Town Planning* London: Routledge.

Bailey, J.T. (1989) *Marketing Cities in the 1980s and Beyond: New Patterns, New Pressures, New Promises* Cleveland, OH: American Economic Development Council.

Barke, M. and Harrop, K. (1994) 'Selling the industrial town: identity, image and illusion' in Gold, J.R. and Ward, S.V. (eds) *Place Promotion: The Use of Publicity and Marketing to Sell Towns and Regions* Chichester: Wiley.

Bassett, K. and Short, J.R. (1989) 'Development and diversity in urban geography' in Gregory, D. and Walford, R. (eds) *Horizons in Human Geography* Macmillan: London.

Beazley, M., Loftman, P. and Nevin, B. (1995) 'Community resistance and mega-project development: an international perspective'. Paper presented to the British Sociology Association Annual Conference, University of Leicester.

Beckett, A. (1994) 'Take a walk on the safe side' *The Independent on Sunday Review* (27/2/94): 10–12.

Bianchini, F. and Schwengal, H. (1991) 'Re-imagining the city' in Corner, S. and Harvey, J. (eds) *Enterprise and Heritage: Cross Currents of National Culture* London: Routledge.

Bianchini, F., Dawson, J. and Evans, R. (1992) 'Flagship projects in urban regeneration' in Healey, P., Davoudi, S., O'Toole, M., Tavsanoglu, S. and Usher, D. (eds) *Rebuilding the City: Property-Led Urban Regeneration* London: Spon.

Bleeker, S. (1994) 'Towards the virtual corporation' *The Futurist* March–April: 11–14.

Boyle, M. and Hughes, G. (1995) 'The politics of urban entrepreneurialism in Glasgow' *Geoforum* 25, 4: 453–70.

Breeheny, M. and Hall, P. (1996) 'Four million households – where will they go?' *Town and Country Planning* (Feb.): 39–41.

Brownhill, S. (1990) *Developing London's Docklands: Another Great Planning Disaster?* London: Paul Chapman.

Byrne, D. (1992) 'The city' in Cloke, P. (ed.) *Policy and Change in Thatcher's Britain*, Oxford: Pergamon.

Cameron, G.C. (1980) 'The economies of the conurbations' in Cameron, G.C. (ed.) *The Future of the British Conurbations* London: Longman.

Castells, M. (1977) *The Urban Question: A Marxist Approach* London: Edward Arnold.

Castells, M. (1983) *The City and the Grassroots* London: Edward Arnold.

Champion, A.G. and Townsend, A.R. (1990) *Contemporary Britain: A Geographical Perspective* London: Edward Arnold.

Christopherson, S. and Storper, M. (1986) 'The city as studio: the world as back lot: the impact of vertical disintegration on the location of the modern picture industry' *Environment and Planning D: Society and Space* 4, 3: 305–20.

Cloke, P., Philo, C. and Sadler, D. (1991) *Approaching Human Geography: An Introduction to Contemporary Theoretical Debates* London: Paul Chapman.

Coleman, B.I. (ed.) (1973) *The Idea of the City in Nineteenth Century Britain* London: Routledge and Kegan Paul.

Cooke, P. (1990) 'Modern urban theory in question' *Transactions of the Institute of British Geographers* (ns) 15, 3: 331–43.

Cornwell, R. (1995) 'Joke City of the rust belt reborn in steel and glass' *The Independent* (11/10/95): 13.

Cox, K. and Mair, A. (1988) 'Locality and community in the politics of local economic development' *Annals of the Association of American Geographers* 78, 2: 307–25.

Crilley, D. (1992) 'Remaking the image of the Docklands' in Ogden, P. (ed.) *London Docklands: The Challenge of Development* Cambridge: Cambridge University Press.

Crilley, D. (1993) 'Architecture as advertising: constructing the image of redevelopment' in Kearns, G. and Philo, C. (eds) *Selling Places: The City as Cultural Capital, Past and Present* Oxford: Pergamon.

Cusick, J. (1990) 'Refuseniks attack "facsimile city"' *The Independent* (31/12/90): 5.

Davis, M. (1990) *City of Quartz: Excavating the Future in Los Angeles* London: Verso.

Davoudi, S. (1995) 'City Challenge: the three-way partnership' *Planning Practice and Research* 10, 3/4: 333–44.

Dicken, P. (1986) *Global Shift: Industrial Change in a Turbulent World* London: Harper and Row.

Duffy, H. (1990) 'The squeeze is on' *Financial Times* Section III: Relocation Survey (26/4/90): 1

Dunn, P. and Leeson, L. (1993) 'The Art of Change in Docklands' in Bird, J., Curtis, B., Putnam, T., Robertson, G. and Tickner, L. (eds) *Mapping the Futures: Local Cultures, Global Change* London: Routledge.

Eyles, J. (1987) 'Housing advertising as signs: locality creation and meaning-systems' *Geografiska Annaler* 69B, 2: 93–105.

Fincher, R. (1992) 'Urban geography in the 1990s' in Rogers, A., Viles, H. and Goudie, A. (eds) *The Student's Companion to Geography* Oxford: Blackwell.

Fretter, A.D. (1993) 'Place marketing: a local authority perspective' in Kearns, G. and Philo, C. (eds) *Selling Places: The City as Cultural Capital, Past and Present* Oxford: Pergamon.

Garreau, J. (1991) *Edge City: Life on the New Frontier* New York: Doubleday.

Glancy, J. (1995) 'Earth Centre awakens revival' *The Independent* (17/5/95): 2.

Glancy, J. (1997) 'Heaven and Earth' *The Independent Magazine* (18/1/97): 26–30.

Glass, R. (1989) *Clichés of Urban Doom and Other Essays* Oxford: Blackwell.

Gold, J.R. and Gold, M. (1990) '"A place of delightful prospects": promotional imagery and the selling of suburbia' in Zonn, L. (ed.) *Place Images in Media* Savage, MD: Rowman and Littlefield.

Gold, J.R. and Gold, M. (1994) '"Home at last": building societies, home ownership and the imagery of English suburban promotion in the interwar years' in Gold, J.R. and Ward, S.V. (eds) *Place Promotion: The Use of Publicity and Marketing to Sell Towns and Regions* Chichester: Wiley.

Gold, J.R. and Ward, S.V. (eds) (1994) *Place Promotion: The Use of Publicity and Marketing to Sell Towns and Regions* Chichester: Wiley.

Goodwin, M. (1992) 'The changing local state' in Cloke, P. (ed.) *Policy and Change in Thatcher's Britain* Oxford: Pergamon.

Graham, S. and Marvin, S. (1996) *Telecommunications and the City: Electronic Spaces, Urban Places* London: Routledge.

Greed, C. (1993) *Introducing Town Planning* Harlow: Longman.

Guskind, R. (1987) 'Bringing Madison Avenue to Main Street' *Planning* (Feb.): 4–10.

Haider, D. (1992) 'Place wars: new realities of the 1990s' *Economic Development Quarterly* 6, 2: 127–34.

Hall, P. (1995) 'Urban stress, creative tension' *The Independent* (21/2/95): 15.

Hall, T. and Hubbard, P. (1996) 'The entrepreneurial city: new urban politics, new urban geographies?' *Progress in Human Geography* 20, 2: 153–74.

Hambleton, R. (1991) 'American dreams, urban realities' *The Planner* 77, 23: 6–9.

Hambleton, R. (1995a) 'The Clinton policy for cities: a transatlantic assessment' *Planning Practice and Research* 10, 3/4: 359–77.

Hambleton, R. (1995b) 'Inner cities: not on EZ street' *Guardian (G2)* (15/11/95): 29.

Hamilton-Fazy, I. (1993) 'Merseyside coverts neighbour's lifestyle' *Financial Times* (26/2/93): 19.

Hamnett, C. (1991) 'The blind men and the elephant: the explanation of gentrification' *Transaction of the Institute of British Geographers* (ns) 16, 2: 173–89.

Hamnett, C. (1995) 'Controlling space: global cities' in Allen, J. and Hamnett, C. (eds) *A Shrinking World: Global Unevenness and Inequality* Oxford: Oxford University Press.

Harris, B. (1995) 'The nature of sustainable urban development' in Brotchie, J., Batty, M., Blakely, E., Hall, P. and Newton, P. (eds) *Cities in Competition: Productive and Competitive Cities for the Twenty-First Century* Melbourne: Longman Australia.

Hartshorn, T. and Muller, P.O. (1989) 'Suburban downtowns and the transformation of metropolitan Atlanta's business landscape' *Urban Geography* 10, 375–95.

Harvey, D. (1973) *Social Justice and the City* London: Edward Arnold

Harvey, D. (1988) 'Voodoo cities' *New Statesman and Society* (30/9/88): 33–5.

Harvey, D. (1989a) *The Condition of Postmodernity* Oxford: Blackwell.

Harvey, D. (1989b) *The Urban Experience* Oxford: Blackwell.

Healey, M. and Ilberry, B.W. (1990) *Location and Change: Perspectives on Economic Geography* Oxford: Oxford University Press.

Hewison, R. (1987) *The Heritage Industry: Britain in a Climate of Decline* London: Methuen.

Holcomb, B. (1990) *Purveying Places Past and Present* New Brunswick, NJ: Urban Policy Research Working Paper no. 17.

Holcomb, B. (1993) 'Revisioning place: de- and re-constructing the image of the industrial city' in Kearns, G. and Philo, C. (eds) *Selling Places: The City as Cultural Capital, Past and Present* Oxford: Pergamon.

Holcomb, B. (1994) 'City make-overs: marketing the post-industrial, North American city' in Gold, J.R. and Ward, S.V. (eds) *Place Promotion: The Use of Publicity and Marketing to Sell Towns and Regions* Chichester: Wiley.

Hubbard, P.J. (1996) 'Re-imaging the city: the transformation of Birmingham's urban landscape' *Geography* 81, 1: 26–36.

Hudson, R. (1989) 'Yacht havens in a sea of despair' *Times Higher Education Supplement* (20/1/89): 18.

Hudson, R. and Williams, A. (1986) *The United Kingdom* London: Harper and Row.

Imrie, R. and Thomas, H. (eds) (1993a) *British Urban Policy and the Urban Development Corporations* London: Paul Chapman.

Imrie, R. and Thomas, H. (1993b) 'The limits of property-led regeneration' *Environment and Planning C: Government and Policy* 11, 1: 87–102.

Imrie, R., Thomas, H. and Marshall, T. (1995) 'Business organisation, local dependence and the politics of urban renewal in Britain' *Urban Studies* 32, 1: 31–47.

Jackson, P. and Holbrook, B. (1995) 'Multiple meanings: shopping and the cultural politics of identity' *Environment and Planning A* 27, 12: 1913–30.

Jacobs, B.D. (1992) *Fractured Cities: Capitalism, Community and Empowerment in Britain and America* London: Routledge.

Jacobs, J.M. (1992) 'Cultures of the past and urban transformation: the Spitalfields market redevelopment in East London' in Anderson, K. and Gale, F. (eds) *Inventing Places: Studies in Cultural Geography* Melbourne, Australia: Longman.

Jameson, F. (1992) *Postmodernism, or the Cultural logic of Late Capitalism* London: Verso.

Jencks, C. (1984) *The Language of Postmodern Architecture* London: Academy Editions.

Judge, D., Stoker, G. and Wolman, H. (eds) (1995) *Theories of Urban Politics* London: Sage.

Keil, R. (1994) 'Global sprawl: urban form after Fordism?' *Environment and Planning D: Society and Space* 12, 2: 131–6.

Knopp, L. (1987) 'Social theory, social movements and public policy: recent accomplishments of the gay and lesbian movements in Minneapolis, Minnesota' *International Journal of Urban and Regional Research* 11, 2: 243–61.

Knowles, R. and Waring, J. (1976) *Economic and Social Geography Made Simple* London: Heinemann.

Knox, P.L. (1991) 'The restless urban landscape: economic and socio-cultural change and the transformation of Washington D.C.' *Annals of the Association of American Geographers* 81, 2: 181–209.

Knox, P.L. (1992a) 'The packaged landscapes of postsuburban America' in Whitehand, J.W.R. and Larkham, P.J. (eds) *Urban Landscapes: International Perspectives* London: Routledge.

Knox, P.L. (1992b) 'Suburbia by stealth' *Geographical Magazine* (Aug.): 26–9.

Knox, P.L. (ed.) (1993) *The Restless Urban Landscape* Englewood Cliffs, NJ: Prentice-Hall.

Knox, P.L. (1995) 'World cities and the organisation of global space' in Johnston, R.J., Taylor, P.J. and Watts, M.J. (eds) *Geographies of Global Change: Remapping the World in the Late Twentieth Century* Oxford: Blackwell.

Knox, P.L. and Agnew, J. (1994) *The Geography of the World Economy* 2nd edn London: Edward Arnold.

Landry, C. and Bianchini, F. (1995) *The Creative City* London: Comedia/Demos.

Lauria, M. and Knopp, L. (1985) 'Towards an analysis of the role of gay communities in the urban renaissance' *Urban Geography* 6, 2: 152–69.

Lawless, P. (1986) 'Inner urban policy: rhetoric and reality' in Lawless, P.

and Raban, C. (eds) *The Contemporary British City* London: Harper and Row.

Lepkowski, D. (1993) 'Crumbling schools: the shame of a city – report' *Birmingham Post* (10/11/93): 1.

Lewis, P. (1983) 'The galactic metropolis' in Platt, R.H. and Macinko, G. (eds) *Beyond the Urban Fringe* Minneapolis: University of Minnesota Press.

Ley, D. (1983) *A Social Geography of the City* New York: Harper and Row.

Ley, D. (1989) 'Modernism, postmodernism and the struggle for place' in Agnew, J. and Duncan, J. (eds) *The Power of Place* London: Unwin Hyman.

Leyshon, A. (1995) 'Annihilating space? The speed-up of communications' in Allen, J. and Hamnett, C. (eds) *A Shrinking World? Global Unevenness and Inequality* Oxford: Oxford University Press.

Leyshon, A. and Thrift, N. (1994) 'Access to financial services and financial infrastructure withdrawal: problems and policies' *Area* 26, 3: 268–75.

Leyshon, A., Thrift, N. and Daniels, P. (1990) 'The operational development and spatial expansion of large commercial development firms' in Healey, P. and Nao, R. (eds) *Land and Property Development in a Changing Context* Brookfield, VT: Gower.

Lister, D. (1991) 'The transformation of a city: Birmingham' in Fisher, M. and Owen, U. (eds) *Whose Cities?* Harmondsworth: Penguin.

Loftman, P. (1990) *A Tale of Two Cities: Birmingham the Convention and Unequal City. The International Convention Centre and Disadvantaged Groups* Birmingham: Birmingham Polytechnic, Faculty of the Built Environment.

Loftman, P. and Nevin, B. (1992) *Urban Regeneration and Social Equity: A Case Study of Birmingham 1986–1992* Birmingham: University of Central England, School of Planning.

Loftman, P. and Nevin, B. (1994) 'Prestige project developments: economic renaissance or economic myth?' *Local Economy* 11, 4: 307–25.

Logan, J.R. and Molotch, H.L. (1987) *Urban Fortunes: The Political Economy of Place* Berkeley, CA: University of California Press.

McCarthy, J. (1995) 'Detroit's Empowerment Zone' *Town and Country Planning* (Dec.): 346–8.

McCracken, G. (1988) *Culture and Consumption: New Approaches to the Symbolic Character of Consumer Goods and Activities* Bloomington: Indiana University Press.

McDowell, L. and Massey, D. (1984) 'A woman's place' in Massey, D. and Allen, J. (eds) *Geography Matters: A Reader* Cambridge: Cambridge University Press.

Malecki, E. (1991) *Technology and Economic Development: The Dynamics of Local, Regional and National Change* London: Longman.

Markusen, A. (1983) 'High tech jobs, markets and economic development prospects' *Built Environment* 9, 1: 18–28.

Massey, D. (1984) *Spatial Divisions of Labour* London: Macmillan.

Massey, D. (1988) 'What's happening to UK manufacturing?' in Allen, J. and Massey, D. (eds) *The Economy in Question* London: Sage.

Massey, D. and Meegan, R. (1982) *The Anatomy of Job Loss: The How, Why and Where of Employment Decline* London: Macmillan.

Mayne, A. (1993) *The Imagined Slum: Newspaper Representation in Three Cities* Leicester: Leicester University Press.

Montgomery, J. (1994) 'The evening economy of cities' *Town and Country Planning* (Nov.): 302–7.

Moudon, A.V. (1992) 'The evolution of twentieth century residential forms: an American case study' in Whitehand, J.W.R. and Larkham, P.J. (eds) *Urban Landscapes: International Perspectives* London: Routledge.

Norcliffe, G.B. and Hoare, A.G. (1982) 'Enterprise Zone policy for the inner city: a review and preliminary assessment' *Area* 14, 4: 265–74.

Oatley, N. (1993) 'Realising the potential for urban policy: the case of Bristol Development Corporation', in Imrie, R. and Thomas, H. (eds) (1993) *British Urban Policy and the Urban Development Corporations* London: Paul Chapman.

O'Connor, K. (1991) 'Creativity and metropolitan development: a study of media and advertising in Australia' *Australian Journal of Regional Studies* (Dec.): 1–14.

O'Connor, K. and Edgington, D. (1991) 'Provider services and metropolitan development in Australia' in Daniels, P. (ed.) *Services and Metropolitan Development: International Perspectives* London: Routledge.

Page, S. (1995) *Urban Tourism* London: Routledge.

Painter, J. (1995) *Politics, Geography and Political Geography: A Critical Perspective* London: Arnold.

Raban, J. (1986) *Old Glory* London: Picador.

Reid, L. and Smith, N. (1993) 'John Wayne meets Donald Trump: The Lower East Side as wild wild west' in Kearns, G. and Philo, C. (eds) *Selling Places: The City as Cultural Capital, Past and Present* Oxford: Pergamon.

Relph, E. (1976) *Place and Placelessness* London: Pion.

Rex, J. and Moore, R. (1967) *Race, Community and Conflict: A Study of Sparkbrook* Harmondsworth: Penguin.

Rex, J. and Tomlinson, S. (1979) *Colonial Immigrants in a British City* London: Routledge and Kegan Paul.

Riddell, P. (1989) *The Thatcher Decade* Oxford: Blackwell.

Rose, D. (1989) 'A feminist perspective of employment restructuring and gentrification: the case of Montreal' in Wolch, J. and Dear, M. (eds) *The Power of Geography* London: Unwin Hyman.

Rose, G. (1992) 'Local resistance to the LDDC: community attitudes and action' in Ogden, P. (ed.) *London Docklands: The Challenge of Development* Cambridge: Cambridge University Press.

Rowley, G. (1994) 'The Cardiff Bay Development Corporation: urban regeneration, local economy and community' *Geoforum* 25, 3: 265–84.

Ryan, K.B. (1990) 'The "official" image of Australia' in Zonn, L. (ed.) *Place Images in Media* Savage, MD: Rowman and Littlefield.

Samuel, R. (1994) *Theatres of Memory: Past and Present in Contemporary Culture* London: Verso.

Sassen, S. (1991) *The Global City: New York, London, Tokyo* Princeton, NJ: Princeton University Press.

Sassen, S. (1994) *Cities in a World Economy* Thousand Oaks, CA: Pine Forge Press.

Saunders, P. (1990) *A Nation of Homeowners* London: Unwin Hyman.

Savage, M. and Warde, A. (1993) *Urban Sociology, Capitalism and Modernity* London: Macmillan/British Sociological Association.

Saxenian, A.L. (1985) 'Silicon Valley and route 128: regional prototypes or historical exceptions' in Castells, M. (ed.) *High Technology, Space and Society* Beverly Hills, CA: Sage.

Scott, A. (1988) *Metropolis: From the Division of Labour to Urban Form* Berkeley, CA: University of California Press.

Shiner, P. (1995) 'Urban regeneration: making less seem more' *Guardian (G2)* (11/1/95): 39.

Short, J.R. (1984) *An Introduction to Urban Geography* London: Routledge and Kegan Paul.

Short, J.R. (1989) 'Yuppies, yuffies and the new urban order' *Transactions of the Institute of British Geographers* (ns) 14, 2: 173–88.

Short, J.R., Benton, L.M., Luce, W.B. and Walton, J. (1993) 'Reconstructing the image of the industrial city' *Annals of the Association of American Geographers* 83, 2: 207–24.

Smith, D.J. (1977) *Racial Disadvantage in Britain* Harmondsworth: Penguin.

Smith, N. and Williams, P. (eds) (1986) *Gentrification of the City* Boston, MA: Unwin Hyman.

Soja, E.W. (1989) *Postmodern Geographies: The Reassertion of Space in Critical Social Theory* London: Verso.

Soja, E.W. (1995) 'Postmodern urbanization: the six restructurings of Los Angeles' in Watson, S. and Gibson, K. (eds) *Postmodern Cities and Spaces* Oxford: Blackwell.

Soja, E.W. (1996) *Thirdspace: Journeys to Los Angeles and Other Real and Imagined Places* Oxford: Blackwell.

Spencer, K., Taylor, A., Smith, B., Mawson, J., Flynn, N. and Batley, R. (1986) *Crisis in the Industrial Heartland: A Study of the West Midlands* Oxford: Clarendon Press.

Stimson, R.J. (1995) 'Processes of globalisation, economic restructuring and the emergence of a new space economy of cities and regions in Australia' in Brotchie, J., Batty, M., Blakely, E., Hall, P. and Newton, P. (eds) *Cities in*

Competition: Productive and Competitive Cities for the Twenty-First Century Melbourne: Longman Australia.

Stoker, G. (1995) 'Regime theory and urban politics' in Judge, D., Stoker, G. and Wolman, H. (eds) *Theories of Urban Politics* London: Sage.

Stoker, G. and Mossberger, K (1994) 'Urban regime theory in comparative perspective' *Environment and Planning C: Government and Policy* 12, 2: 195–212.

Stone, C. (1989) *Regime Politics: Governing Atlanta, 1946–1988* Lawrence: University Press of Kansas.

Stone, C. (1993) 'Urban regimes and the capacity to govern: a political economy approach' *Journal of Urban Affairs* 15, 1: 1–28.

Strassman, W.P. (1988) 'The United States' in Strassman, W.P. and Wells, J. (eds) *The Global Construction Industry* London: Unwin Hyman.

Sudjic, D. (1993) *The 100 Mile City* London: Flamingo.

Taylor, E. (1991) 'It's no carnival when the city steals the show' *The Independent* (5/9/91): 25.

Teather, E.K. (1991) 'Visions and realities: images of early postwar Australia' *Transactions of the Institute of British Geographers* (ns) 16, 4: 470–83.

The Economist (1995) 'From screwdrivers to science' (18/3/95): 32–4.

Thomas, H. (1994) 'The local press and urban renewal: a South Wales case study' *International Journal of Urban and Regional Studies* 18, 2: 315–33.

Thrift, N. (1987) 'The fixers: the urban geography of international commercial capital' in Henderson, J. and Castells, M. (eds) *Global Restructuring and Territorial Development* London: Sage.

Turok, I. (1992) 'Property-led regeneration: panacea or placebo?' *Environment and Planning A* 24, 3: 361–79.

Turok, I. (1993) 'Inward investment and local linkages: how deeply embedded is Silicon Glen?' *Regional Studies* 27, 5: 401–17.

Ward, S.V. (1988) 'Promoting holiday resorts: a review of early history to 1921' *Planning History* 10, 1: 7–11.

Ward, S.V. (1990) 'Local industrial promotion and development policies 1899–1940' *Local Economy* 5, 2: 100–18.

Ward, S.V. (1994) 'Time and place: key themes in place promotion in the USA, Canada and Britain since 1870' in Gold, J.R. and Ward, S.V. (eds) *Place Promotion: The Use of Publicity and Marketing to Sell Towns and Regions* Chichester: Wiley.

Watson, S. (1991) 'Gilding the smokestacks: the new symbolic representations of deindustrialised regions' *Environment and Planning D: Society and Space* 9, 1: 59–71.

Whitehand, J.W.R. (1990) 'Makers of the residential townscape: conflict and change in outer London' *Transactions of the Institute of British Geographers* (ns) 15, 1: 87–101.

Whitehand, J.W.R. (1994) 'Development cycles and urban landscapes' *Geography* 79, 1: 3–17.

Whitehand, J.W.R. and Larkham, P.J. (eds) (1992) *Urban Landscapes: International Perspectives* London: Routledge.

Whitehand, J.W.R., Larkham, P.J. and Jones, A.N. (1992) 'The changing suburban landscape in post-war England' in Whitehand, J.W.R. and Larkham, P.J. (eds) *Urban Landscapes: International Perspectives* London: Routledge.

Wolman, H.L., Ford, C.C. III, and Hill, E. (1994) 'Evaluating the success of urban success stories' *Urban Studies* 31, 6: 835–50.

Wynn Davies, P. (1992) 'Development losses of £67m "a scandal"' *The Independent* (13/7/92): 2.

Zukin, S. (1988) *Loft Living: Culture and Capital in Urban Change* London: Radius.

Index